*Understanding the Eastern Caribbean and the Antilles*

Understanding the Eastern Caribbean and the Antilles

with Checklists Appended

Nelson Marshall

**Professor Emeritus
of Oceanography and
Marine Affairs
University of Rhode Island**

Th'Anchorage Publisher

Copyright © 1992 by Nelson Marshall

All rights reserved. Except for brief quotations in critical articles and reviews, no part of this book may be used or reproduced in any manner whatsoever without written permission from the author, care of: Th'Anchorage Publisher, P.O. Box 1056, St. Michaels, Maryland, 21663, U.S.A.

First Printing 1992

Cataloging-in-Publication Data
Marshall, Nelson
Understanding the Eastern Caribbean and the Antilles, with Checklists Appended.
Subject area: Natural history—Caribbean area, Man—Influence of nature—Caribbean area, Ecology—Caribbean area, Marine biology—Caribbean area.

ISBN 0-9628730-0-4

Printed in the United States of America

Dedicated to Archie Carr, 1909–1987

In appreciation of his ability to narrate his insights into Caribbean natural history. No other teacher has influenced me more.

Contents

Acknowledgments ix
Foreword xiii

Introduction 1
The Island Arc 3
The Surrounding Sea 10
Seen at Sea 18

Coral Reefs: Characteristics and Viewing 24
Atolls in the Caribbean—Yes or No? 35
Mangroves and Mangrove Swamps 39
Seagrasses 42
The Sandy and Rocky Shore Environment 46
Beachcombing 51

Fisheries 56
Some Unique Fish Behavior 61
Aquaculture: Present (and Future?) 65
Sea Turtles 70
The Queen Conch 74
The Spiny Lobster 79
Whaling in Bequia 82
Living Light in the Sea 86

The Floral Landscape 88
Livestock, Crops, and the Local Market 96
Variety in Nature 101

Wind, Waves, and the Beaufort Scale 114
Hurricane Winds and Seas 122
The Bermuda Triangle—Sheer Nonsense 131
The Tropical Sky 135

The Indies of Columbus 139
History as Influenced by Natural Features 146
The Major Resources 152

Law of the Sea and the Island Countries 154
Conservation 157
A Pollution Wrap-up 163

Observing and Photographing 167
Beware 173

Appendices
  Introduction 185
  Some Useful Field Guides 188
Checklists
  Fishes (Plus Sea Turtles) 192
  Corals, Gorgonians and Anemones 202
  Shells 205
  Marine Invertebrates 217
  Marine Bottom Plants 223
  Birds 227
  Flowers, Shrubs, and Vines 233
  Trees 244
  Pages for Added Observations 252

Acknowledgments*

When you consider the scope of this book you will realize that I needed a great deal of help to cover the information presented. Especially helpful were the inputs of Bill Dennison, an enthusiastic scientist-naturalist colleague at the Horn Point Environmental Lab of the University of Maryland. I am also especially appreciative of the work of Debbie Kennedy, the talented artist who produced most of the drawings.

A series of experts reviewed and improved upon the appended checklists. Specifically, C. Lavett "Smitty" Smith of the American Museum of Natural History checked the fishes. Bob Shoop of the University of Rhode Island and Peter Pritchard of the Florida Audubon Society answered inquiries relative to turtles. Walter Sage, also of the American Museum, helped with the shells. Alina Szymant of the University of Miami reviewed the corals and marine invertebrates. Eleanor Gibney, horticulturist for Caneel Bay, Inc. in St. John, Virgin Islands, and Gary Ray, Ph.D. candidate doing research in St. John, helped with the terrestrial flora.

In addition to the inputs of those named above plus those who are credited with the various illustrations, many of whom also helped with the text, I was in touch with countless others. Adequate coverage

---

*My informality in listing Bill, Debbie, Smitty, etc. stems from my assumption that those who take the trouble to read these Acknowledgments are trying to spot the names of their friends and acquaintances and know them best by first names and nicknames. Many who are listed have impressive titles, such as Ph.D., Professor, Chief or Captain, but I doubt that any of them will feel slighted if we skip such recognition.

Acknowledgments

seems impossible, but I will try. Colleagues from the University of Rhode Island with useful inputs include three of my former students, Hank Parker at Southeastern Massachusetts University, John Tietjen at City University of New York, and Ray Gerber at St. Joseph's College. I am also grateful for help from Jim Parrish and Phil Richardson (both ex URI students, the former now located at the University of Hawaii, the latter at the Woods Hole Oceanographic Institution), Prentice Stout, Mike Pilson, Howard Winn, Lew Alexander, Virginia Tippie (another URI alum now with NOAA), Haraldur Sigurdsson, Rich Appledorn (an alum now with the University of Puerto Rico), Jim Griffin, and Daven Joseph (now a marine resources planner for the Organization of Eastern Caribbean States).

Among the many who have been helpful through my associations at the Horn Point Environmental Laboratory I think of Bill Boicourt, Roger Newell, Charles Hocutt, Jeff Cornwell, Jenny Purcell, Judy O'Neil Dennison, and Paul Spitzer. My writing spanned the terms of two directors, Mike Roman and Tom Malone, who, with their assistant Sharon Foxwell, extended numerous courtesies as I served at the lab in the capacity of Adjunct Professor. Darlene Windsor helped obtain interlibrary loans. Jane Gilliard was indispensable as the typist with the uncanny knack of making order out of my scribbling.

Included among the many others I relied on are: David and Joan Robinson of the Alexander Hamilton Museum in Nevis, Miranda Wecker of the Council on Ocean Law, David Lewis of McGill University, Charlie Yentsch of the Bigelow Oceanographic Laboratory, Kathy Orr, an independent writer and artist, Mel Goodwin of the South Carolina Sea Grant College, Ian Macintyre of the Smithsonian Institution, Art Dammann of St. John, Virgin Islands, Bob Lippson of the National Marine Fisheries Service, William Herrnkind of Florida State University, Desmond Nicholson of the Antigua Archeological Society, Milan Keser of the Millstone Environmental Laboratory, Jude Wilber of the Sea Education Association, Judy Lang of the Texas Memorial Museum, Ritchie Bell, emeritus from the University of North Carolina, and Bill Wroten, emeritus from Salisbury State University.

Acknowledgments xi

For background I had addressed a series of questions to officials in charge of departments concerned with conservation on the various islands. For brevity in recognizing those who provided information I will simply name individuals and islands represented. Replies were received from Mark Eckstein (St. Lucia), François vander Haeven and Godfried Richardson (both from St. Maarten), John Kelly (Antigua), Felix Gregoire (Dominica), and Jennifer Cruickshank and B. Johnson (both from St. Vincent). A useful wildlife assessment report was received from Wayne Arendt (USDA Forestry Service in Puerto Rico) and John Faaborg (University of Missouri, Columbia).

For my travels into the Eastern Caribbean on board various seagoing craft I am indebted to the scientific staff and crew involved for each trip. From the University of Rhode Island we took R.V. *Trident*, with Ted Smayda as chief scientist, to Bahia Fosforescente and on to Jamaica where I had the good fortune to discuss Caribbean corals with the late Tom Goreau. Later, as chief scientist, I took *Trident* to the Bahamas reefs with such top-notch coral scientists as Bob Johannes, Phil Helfrich and Bill Wiebe participating. Earlier I had gone far to the south in the Bahamas on the R.V. *Oliver* of the American Museum of Natural History with Lavett "Smitty" Smith as chief scientist.

When I sailed our own boat, *Argo*, to the Virgin Islands, Bob and Barbara Welsh, Jim Parrish, Ann Durbin and Greg Telek helped with studies we made on the reefs. On the second of two trips through the Lesser Antilles on the small cruise ships of the American Canadian Line, Chan and Edie Swallow were along and devoted considerable time going over an early draft of this book.

My two relatively recent trips aboard the brigantines of the Sea Education Association stand out not only because we worked good reef sites but also because I was surrounded by enthusiastic college students whose questions kept me on my toes. Obviously I cannot list all 48 students plus the members of the staff and crew but can at least acknowledge the help of Chuck Lea and Terry Hayward, Chief Scientist and Master on the first trip, and Paul Joyce and Greg Lohse, Chief Scientist and Master on the second. I'll add a special note of recognition to Bill Zuber, a SEA ship engineer, who was a great help both to the student group and to me.

Finally I had the much needed assistance of two good readers, namely Howard Bloomfield, an independent author and an "old pro" at proofreading, and Grace Terry Marshall, my wife, who reviewed my writing style and made certain I avoided unfamiliar jargon.

I am sure that, as she reads this and recalls the countless inquiries she overheard from my phone conversations, Grace will wonder how my acknowledgments can help but fall short of the total. No doubt there are serious omissions. At least I can conclude with a word as to how gratifying it has been to have the help of so many. Falling back on fifty years in marine and related studies, I could always find someone, sometimes a "long-lost old friend," at the "other end of the phone" who could answer my varied and often difficult questions.

 *Foreword*

The Caribbean Sea has a spectacular variety of islands that are increasingly accessible to North Americans and Europeans in search of sunshine and beaches. But beyond the beaches, the Caribbean isles and turquoise seas have a rich natural history. In *Understanding the Eastern Caribbean and the Antilles,* Nelson Marshall has provided delightful and entertaining insights into the natural history of the Caribbean that will help visitors become travelers and explorers, instead of tourists simply seeking sun and sand.

There are few people who could actually pull together the various vignettes collected in this work. Only a well-traveled natural historian with connections in Caribbean marine science could unveil such a variety of natural history treasures. And, luckily for us, Professor Marshall is such a person. His training in the formative years of American oceanography provided him with a broad natural history background. The result of his expertise and continued interest in the Caribbean culminated in this series of short descriptions of a variety of natural history features. In addition, Professor Marshall has delved into many of the social issues affecting the Caribbean environments. The observations in *Understanding the Eastern Caribbean and the Antilles* are described in non-technical terms with an easy-to-understand style. Yet these are distillations of complex processes that Professor Marshall has formulated based on rigorous research to insure that he would "get it right."

This collection of short chapters on a diversity of subjects may stimulate readers to learn more about aspects of Caribbean natural history. The checklists and field guides referred to in the appendix provide a starting point for continued learning. But most importantly, this book should provide readers with an appreciation of the natural history wonders of the Caribbean region and allow them to become transformed into travelers with an eye for the world around them.

William C. Dennison
Research Associate Professor
Horn Point Environmental Laboratory
Cambridge, Maryland
July 1991

Understanding the Eastern Caribbean and the Antilles

 *Introduction*

It has been my good fortune to visit the tropics many times in connection with research projects on coral reefs. While enjoying the spectacular panorama of life underwater and ashore, I have also been frustrated, trying to comprehend the diversity of my surroundings. Good field guides are very useful, and checklists as appended herewith are an important crutch to observing.

As I developed such lists for the Eastern Caribbean it also became apparent that an understanding of the region would be enhanced by concise notes on its natural history. As the notes unfolded and as it seemed they might be useful to others, they have been presented in this handy book. My objective is to offer a series of brief write-ups of the natural features of the Eastern Caribbean that would require a long bookshelf of more detailed readings if the topics were to be dealt with in depth. For some this orientation may suffice, at least for starts. In addition those pursuing more specific interests, and even experts and specialists, may welcome the background provided by the diversity of subjects covered. Though intended primarily for travelers, this may be welcomed by residents as well. While the information contained focuses on the islands of the Eastern Caribbean, the accounts should also be useful in the Bahamas and along the Florida Keys.

The opening write-up of "The Island Arc" of the Greater and

Introduction

Lesser Antilles, sets the stage for all that follows. Similarly the next section, "The Surrounding Sea," helps to orient the reader to the environment. Sensing that some will be sailing or will be aboard ship as they read this, the notes move immediately to observations to be made while underway. The accounts that follow consider the coral reefs, the fisheries, the floral landscape, conservation, law of the sea applications, and much more. And there are various *do-it-yourself* challenges with pages for your own notes plus blanks for recording further entries in the checklists.

 # The Island Arc

East of the Caribbean there is a subduction zone where the Atlantic Plate* of the earth's crust is being pushed downward, "slipping under" the Caribbean Plate. The resulting instability beneath the ocean floor has caused volcanoes to rise from the deep. The rise forms the Caribbean Island Arc, which has an older axis to the east and a newer axis to the west. Simplified, the process may be described diagrammatically as follows:

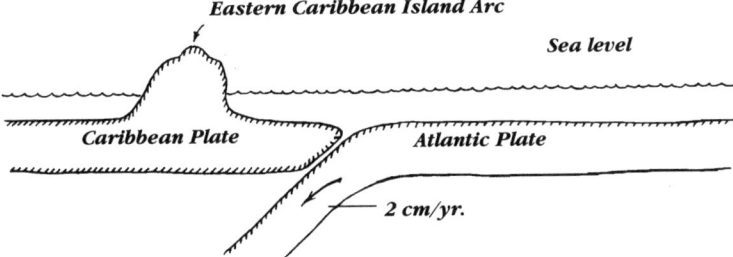

On bathymetric charts and maps the dip in the ocean floor where the plates converge shows up to the east of the islands, arching around to the northwest. To the south a pile-up of sediment where the plates collide accounts for the island of Barba-

*This terminology is a simplified way of referring to the combined North and South American Plates being subducted in this region.

dos. On the other hand, north of Puerto Rico there is very little sediment filling the subduction region and the resulting Puerto Rican Trench of more than 8,200 meters is the deepest area of the Atlantic Ocean.

The volcanoes of the more western axis of the Arc are young in terms of the geologic time scale and a few remain active: Martinique's Mt. Pelee erupted in 1902, killing nearly 30,000 people and La Soufrière on Saint Vincent erupted as recently as 1979. The more eastern axis of the Arc is older and has been subject to erosion over a longer time period; therefore it is not as mountainous. In fact, to an appreciable extent, the terrain to the east is overlaid with marine formations. Guadeloupe, with its two distinct regions, nicely illustrates the two axes, having a mountainous terrain to the west and a relatively low lying sector to the east.

The role of volcanoes in the region is dramatically illustrated by "Kick-em Jenny," which is the most active volcano in the Arc but is yet to form an island.* The crater, north of Grenada and just west of Ronde, is more than 1,000 meters above the deep sea floor but still about 150 meters from the ocean surface. It erupted in December, 1988 with gases and ash breaking the sea surface. Since Kick-em Jenny is on the route from the Grenadines to St. George's Harbor in Grenada, it can pose a threat to shipping and boating, and there are times when traffic has been warned to stay clear. My colleague, Haraldur Sigurdsson of the University of Rhode Island, has kept close watch on Kick-em

*Charts also show "Kick-em Jenny" as an alternate name for nearby Diamond Island. When the submerged volcano breaks the surface, if not before, there will be a need to have separate names to avoid confusion.

The Island Arc 5

Jenny. He suggests that, at its present rate of growth, it will emerge as a volcanic island early in the next century.

Inevitably the volcanic terrain that is so impressive in the western islands is hilly and often very rugged. Most striking are the poster-perfect Pitons on the west coast of St. Lucia, where the craters have eroded away but the resistant cores of two volcanoes of the past persist. The sulfur springs nearby, known as soufrières with their boiling water, steam, and the smell of sulfur, serve as striking evidence that the volcanism is still active.

Underwater view of the rim of Kick-em Jenny, a volcano growing toward the surface. (Photo in 1988 by Haraldur Sigurdsson, University of Rhode Island.)

Flatlands are common along the older eastern axis of the Arc, but they are limited to the periphery of the more mountainous western islands. The various low-lying areas may be outwash plains from eroding hillsides, or they may be marine accretions behind barrier reefs and lagoons. Due to the sporadic nature of volcanic intrusions, accompanied by uplift and subsidence, and due to changing sea levels in earlier geological times, complex combinations of volcanic and marine rock and soil are to be found. In a few places where the low flat areas are frequently covered by shallow films of seawater, deposits of salt, sometimes managed for their commercial value, form from the evaporation of invading seawater.

Cities, towns and villages tend to develop on the low-lying ground or to cluster low on the hillsides of these Caribbean islands. Agriculture, however, is practiced on the hillsides as well as on lower ground. The materials that accumulate from the erosion of old volcanic outpourings, coupled with seasonal high rainfall, serve as good cropland.

On some islands one sees a marked difference between the wetter slopes exposed to trade winds from the east and the relatively dry, but seldom barren, slopes on the lee side. Tropical rainforest conditions sometimes develop in the highlands, especially where the steep windward slopes are high enough to force a rise of air that is sufficiently rapid to cause the cooling clouds to "dump their water." Thus El Yunque, the tropical rainforest of Puerto Rico, is at the eastern end of the island while the west is a drier area where sugar cane is grown.

The volcanic origins of the Island Arc account for the dark sand beaches that one may encounter. Though no cleaner, the preferred white sand* beaches derived from the carbonate deposition in the surrounding sea, as discussed in the next section, are more common. Along these white sand beaches one occasionally encounters slabs where the beach sediments have become compacted and tightly bound to form so-called "beach rock." The formation of such rock gives one a sense of how sands, reefs and marine accretions in general form limestone bedrock in this region. When limestone of varied consis-

---

*Visitors from temperate climates are accustomed to sand made up largely of ground bedrock, thus having the more yellow appearance that we often refer to as "sand" color.

tency is partially dissolved and subjected to some biological degradation, the result can be a very rough surface, almost impossible to walk on. Where the solution of the limestone has resulted in substantial depressions or sinks, sometimes dropping into underground solution channels, one sees what is known as karst topography. One of the largest radio telescopes has been constructed in the large karst sinkhole near Arecibo, Puerto Rico.

Look along the edges of the islands, particularly on the elevations to the west, for indicators of their geological history. Layers of sedimentary rock may be evident, sometimes with a dip where the formation has shifted. Conspicuous layering may relate to the sporadic volcanic intrusions or changes in sea level, perhaps both. The rock face may be pitted from erosion of the volcanic rock, the solution of limestone, or the irregular resistance to such forces where marine and volcanic material are intermixed.

Where freshwater runs off the high terrain, excessive silt loads and pollution may occur. These adverse conditions induced by man are discussed further in the write-ups of "Coral Reefs" and "Pollution."

No doubt you have heard the expressions "Greater and Lesser Antilles," also "Windward and Leeward Islands." Of the Greater Antilles, Puerto Rico is the only one specifically covered in this book. Hispaniola, Cuba and Jamaica (not discussed herein) lie to the west. The smaller islands to the east are collectively referred to as the Lesser Antilles. Since a sailing vessel approaching on the prevailing winds from across the Atlantic is likely to make a landfall somewhere from Martinique to the south, the southern group of the Lesser Antilles is referred

The Pitons of St. Lucia. (Courtesy of Allan Smith, Caribbean Natural Resources Institute.)

to as the Windwards. The rest, in other words those to the north and west, are called the Leewards.

The following islands along the eastern part of the Arc and formerly colonies of the United Kingdom are now independent:

> St. Kitts-Nevis
> Antigua and Barbuda
> Dominica
> St. Lucia
> St. Vincent and the Grenadines
> Grenada
> Barbados
> Trinidad and Tobago *(east of South America and not always regarded as part of the Lesser Antilles)*

Puerto Rico is a Commonwealth of the United States. St. Thomas, St. John, St. Croix and neighboring small islands make up the United States Virgin Islands. Tortola is the main island of the British Virgin Islands. The United Kingdom possessions extend to the east to include Anguilla and Montserrat. Martinique is a department of France. The island labelled St. Martin on the map includes two political entities. St. Martin is on the northern half and, along with the island of St. Barthélemy, is linked to Guadeloupe as a French regional government. That half of the island that lies south of St. Martin is St. Maarten. This is Dutch and, with the islands of St. Eustatius and Saba plus Curaçao and Bonaire off the north coast of South America, make up the Netherlands Antilles as an integral part of the Kingdom of the Netherlands.

The Surrounding Sea

The tradewinds, blowing across the Atlantic, almost equally from the northeast and east at these latitudes, are among the forces driving the westerly currents into the Caribbean. This is all part of the westward circulation from across the Atlantic Ocean that ultimately "feeds" the Gulf Stream system. Obviously the currents are strongest where the flow from the east passes between the islands on entering the Caribbean basin. Just north of Grenada, currents occasionally reach 3 knots. Lesser drifts are common between islands that lie more to the north.

Tidal currents ebb and flood at 6-hour intervals. However, even on the dates of spring tides, when the gravitational pull of the moon is greatest, the tidal range is seldom as much as two feet (a little over one-half meter). Accordingly, tidal circulation is not a major factor to consider, though sometimes tidal currents flowing against the prevailing winds and oceanic circulation can increase turbulence, particularly between the islands.

Though the trend of the onward flow into the Caribbean basin is generally westward, there are some large and varied rotating systems or gyres behind the islands. Horizontal and vertical mixing may occur in these systems. Vertical mixing may bring nutrients up from the depths and enhance the productivity of the water column. Horizontal mixing may retain larval stages so that they may settle near where they have been spawned. Perhaps this is the mechanism that explains the retention of larvae that would seem free to drift many miles away as exemplified in the write-up of the "The Queen Conch."

The tradewinds of the region may reach velocities of 30 knots and more, tending to be greatest in the middle of the dry season, at the time of the so-called "Christmas Winds." High winds may result in waves of more than 4 meters or about 15 feet from trough to crest windward of the islands and in other areas where there is no protection from thousands of miles of open ocean. Occasionally high pressure weather systems bearing down on the area as fronts from the north tend to stall the trades, perhaps even resulting in a few days of "norther" winds. By early June, after the sun has moved appreciably to the north and the center of the tradewind belt has shifted northward, rainfall increases appreciably, condensing from the warm air rising over the ocean surface. Thus there is a wet season from some time in June to early November, and a dry season from November to at least through May. During the wet season the tropical storms forming at these latitudes may develop into hurricanes. For more on sea conditions, see "Wind, Waves and the Beaufort Scale" and "Hurricane Winds and Seas."

With the tropical Atlantic as the source, surface water temperatures are in excess of 25°C (around 80°F) and remain so year-round. Areas with even higher water temperatures may develop where intense sunlight heats lagoons and extreme shallows. Surrounded by such warm water, the island temperatures are also warm, yet rarely extremely hot. Since the water is never cold, fog is a rarity, though visibility can become limited in severe downpours. Reflecting the oceanic source, the salt content of the water is close to 35 parts per thousand. Dilution from freshwater runoff does not extend far offshore, but can be considerable around the mouths of estuaries, and in lagoons.

With the increasing temperature and salinities often occurring in Caribbean waters the calcium carbonate, present at supersaturation concentrations, is released from solution. These conditions are "very friendly" to life forms that require the carbonate. Some, like the corals, use the calcium compound for their internal and basal skeletal structure; some, like molluscs, use it for their shells; some, as with certain bottom algae, incorporate calcium carbonate as an integral part of their tissues; and some simply show an external white precipitate. Appropriately it has been said that an environment of this sort is a regular "carbonate factory." One even finds "white muds" in places where carbonate settles onto thick mats of growth on the bottom. Though the corals are the most spectacular products of this environment, they are but part of the broad pattern of carbonate deposition. —See how many different examples of carbonate deposition you can discern. I have even seen white precipitates on sunken tires.

The skeletal fragments of this environment make up the white sand of the beaches. Included are the plates of colonial algae such as halimedas and tiny snail-shaped shells of a group of one-celled animals known as forams which, if abundant, may give the sand a pinkish tinge. Then there is a great variety of fragments from corals, shellfish, etcetera. While much of this is simple calcium carbonate (called limestone when compacted), some is mixed with magnesium and is known as dolomite. If you were to grind almost any shell, taken from anywhere in the world, into sand-sized particles and were to bleach out the organic matter, the end result would be much the same in kind and appearance as the carbonate sand of the tropics. As noted in "The Island Arc," one also sees black or dark gray sands derived from ground-up volcanic rock.

Where seagrasses are abundant in the shallows, their green color may predominate over the reflection off the white sand bottom. Where living coral communities are well established, they reflect as dark brown areas. For fringing or barrier reefs the dark effect tends to run parallel, adjacent to, or off the shoreline; otherwise the coral growths may reflect as scattered patches. When the sun is overhead the dark brown of the coral serves as a warning when navigating such waters. Unfortunately, when the shadows from passing clouds darken the surface, the dark effect from the corals and the dark shadows are not easy to distinguish.

Several factors contribute to the appearance of the water beyond the shallows. Color may result from light reflected off suspended matter, occasionally including a unique abundance of microscopic plant and animal life. However, in these tropical seas where the water is typically very clear, an intense blue color often prevails simply because the longer wave lengths, in other words the reds, yellows and greens, are absorbed and the short, penetrating blue reflects off very fine particles. This effect has been compared to the blue in the sky, the explanation being much the same. The deep blue of such clear, oceanic water is sometimes called the "desert color of the sea".

The various turquoise shades and tints, that are such an impressive feature of waters of intermediate depths in coral reef regions, are not as readily explained. Obviously the penetrating blue contributes but, without depth, this color is "diluted." Reflection off the white sand bottom also contributes but, where a greenish tint is noticeable, additional inputs must be involved. Since various shades of turquoise are seen, it would seem that contributing factors may vary and these are not readily identified. One possibility is a suspension of one-celled plants. Though, as discussed below, these are not likely to be abundant, their presence may be sufficient to tint the water. Another possibility might be the presence of dissolved yellow (humic) organic matter. Some attribute the coastal green water off the continents as a blend of this yellow with the penetrating blue. Though probably not concentrated around the islands and reefs, there may be enough of this yellow to blend and add a tint to the turquoise.

Incidentally, for depths greater than 300 meters even the remnants of blue light are absorbed and there is *total* darkness. More information on color absorption is given in the write-up "Observing and Photographing." As you will see, when considering photography, aggregated organic material looking like and called "marine snow" may appear in the water column.

As suggested by the clarity and the prevalence of the "desert color," Caribbean water is characteristically nutrient-poor with relatively little production. Yet the fundamental relationships within the system are not unlike those of seawater elsewhere. Sparse though it may be, the drifting one-celled plant life, the phytoplankton, is the "grass" or "pasture" of this environment. Slightly larger drifting animals, the zooplankton, feed on this.

The cycles within this system are quite complex, with intermediates linking the one-celled plants and the larger zooplankton in food chains that ultimately lead to larger consumers, including fishes. Many of the corals, shellfish, lobsters, crabs, in fact a large percentage of the bottom dwelling forms, have larval stages that drift about as zooplankton before settling.

An exception to the clarity of the oceanic water entering the Caribbean from the east may occur toward the south where the influence of the Orinoco and Amazon Rivers can be noticeable. Actually the mouth of the Amazon is far to the south but it is the mightiest of all rivers and its plume persists as much as 300 kilometers out to sea. Though much of the silt load of the Orinoco and Amazon has settled out by the time these waters have been entrained into the open-Caribbean, the water may have the greenish tint of coastal waters as contrasted with the oceanic blue.

Obviously you will not see the bottom topography; however, a few comments are in order. In some areas submerged "shoulders" or shelf areas off the islands extend seaward from the shoreline. Elsewhere, as noted in the discussion of the Island Arc, the steep slopes to very deep water are not far off. Tem-

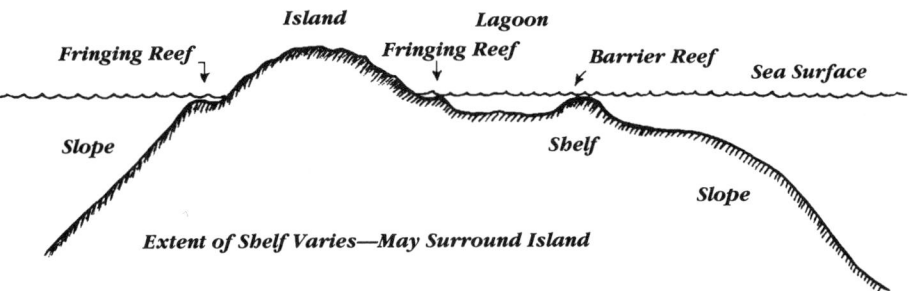

peratures drop markedly with depth and may be as low as 6°C (43°F) at no more than 1,000 meters from the surface.

A process known as Ocean Thermal Energy Conversion (OTEC) involves pumping the deep, cold water to the surface to convert into electricity the energy of the 20-degree bottom-to-surface temperature differential. For Caribbean islands, where deep water is commonly very close to shore, this has been seriously considered, and the interest is enhanced by the fact that one such generating system produces freshwater as a by-product. Also, the deep water is rich in nutrients which, when brought to the surface, might serve to fertilize aquaculture undertakings. OTEC pilot projects have been carried out, particularly on the big island of Hawaii, but as yet there are no commercial applications of this principle.

The thought of low temperatures in the deep basins of the Caribbean points to one of the *fun* questions of physical oceanography. —How does cold water, even lower than 4°C in the extreme, get into the basins of the Caribbean? The very cold, high salinity properties of the deep water in the Atlantic are well known. The general source is not difficult to comprehend for the dense cold water sinks in the arctic regions and drifts very slowly at great depths, to reach the lower latitudes. But it would seem that such dense water could not flow into the Caribbean for there are no deep passes where this might occur. Sills such as in the Jungfern Passage at 1,860 meters south and west of St. Croix, and in the Windward Passage at 1,688 meters between Cuba and Hispaniola, are not deep enough. Direct probing of this barrier to water exchange was pioneered by my former colleague, Tony Sturges, when he made direct bottom measurements at the Jungfern sill using instrumentation lowered from the University of Rhode Island research ship, R.V. *Trident.* Essentially he observed surges of the deep Atlantic water over the sill, in spite of what one expects on the basis of density stratification. The specifics of his observations and his explanations of the hydrography of such sporadic surges are complex and may not be the last word. However, with this phenomenon and exchanges in like manner across the sill of the Windward Passage, the low temperature at depths in the Caribbean is no longer such a mystery.

 *Seen at Sea*

It is not unusual to see bottle-nosed dolphins* (porpoises) frolicking at the bows of ships and yachts underway at sea. Large schools are referred to as "pods." And in the distance you may see the spouts of the small pilot whales or of the giant sperm and humpback whales. Perhaps you will see all three whale species, but probably not together. If the whales are not too far off you may see them breaching or rolling at the surface. The spouts of the two large whales may be distinguished; that of the humpback is bushy, while the blow of the sperm whale angles forward. Approach if you wish, under steady throttle or sail, but no closer than 100 meters (300 feet) to avoid startling them.

You are bound to see flying fish skittering off at an angle to the bow. Incidentally, they don't fly but glide, perhaps getting an added thrust occasionally from the lower lobe of the tail fin. You might drum up a little competition to see who can spot the longest glide. Sometimes the flying fish land on deck at night. Save them; they are good to eat.

If you keep a sharp lookout you may see large sea turtles. Identifying them at a glance is a bit unlikely.

Sporadically the sea-blue floats of the Portuguese man-o-war

*The dolphins referred to here are mammals, relatives of whales. There is also a pelagic fish known as a dolphin.

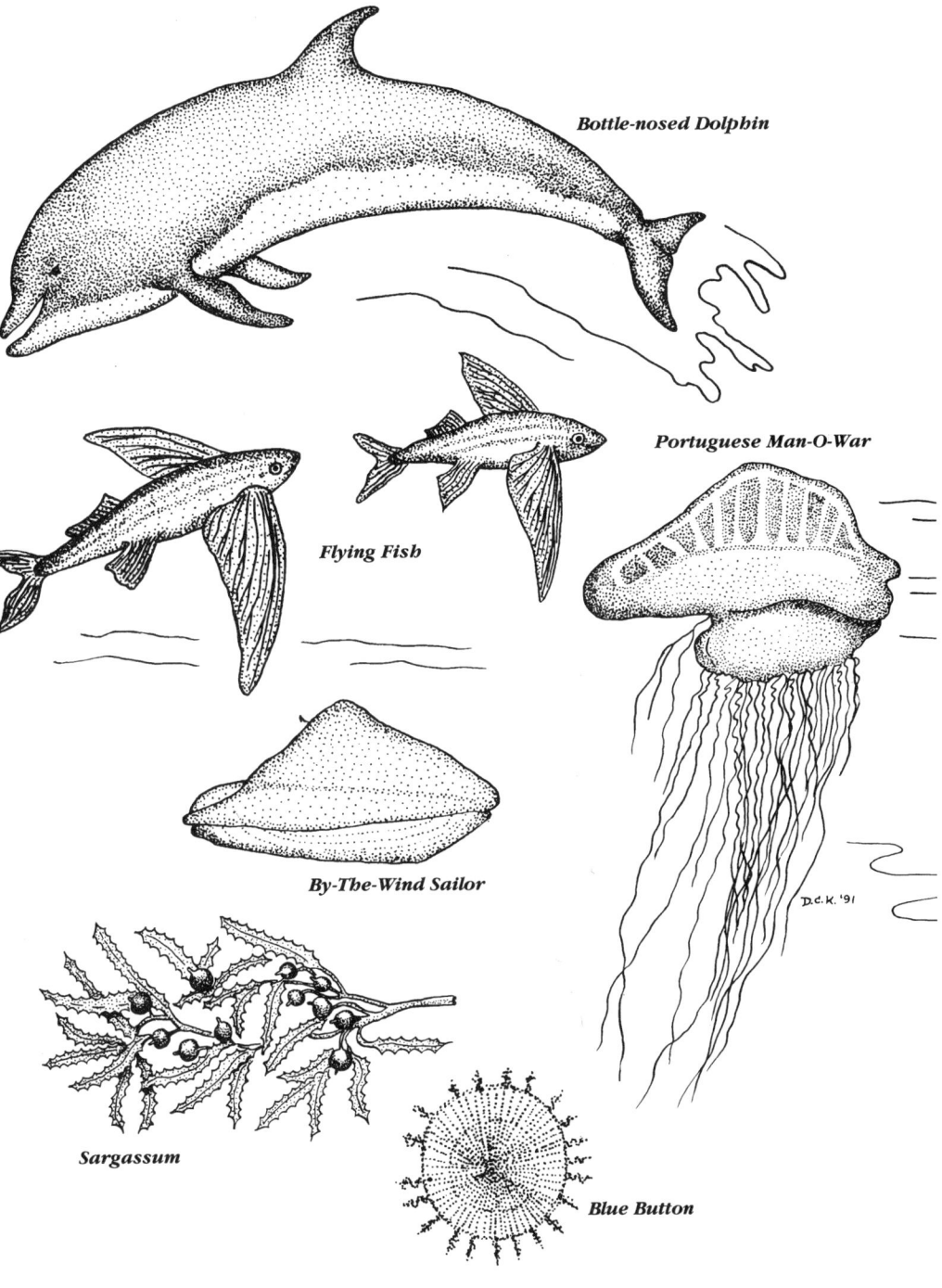

are rather common on the ocean surface. Beneath the float there are "strings" of reproductive and feeding bodies with cells that can inflict a very severe sting. A little blue fish lives under the float, enjoying immunity to the stingers and feasting on plankton that are not so fortunate. Also on the surface is a small, blue jellyfish with an upright "sail." This is known as *Vellela* or by-the-wind-sailor. Rounding out this group of "floaters" is the tiny, one-inch across, blue button. True to the name, it does look like a blue button. Finally, at the surface you may see small clusters of purple sea snails, *Janthina*, drifting in the bubble floats they secrete, but you are more likely to collect their shells that wash up on the beach. A bit deeper, you may see the round moon jelly and possibly some of the many other jelly-like forms so important in the ecology of the oceans.

The floats of the man-o-war and the "sails" of the *Vellela* catch the wind which moves them across the sea. In some individuals the wind-catching surface slopes to the right; in others to the left, causing them to drift away in different directions. Some biologists have speculated that this is an adaptation for survival, with some safely sailing out to sea while others are blown ashore. In a lighter vein some of my friends have suggested that perhaps the "righties" are one sex, the "lefties" the opposite, with the winds bringing them together for happy mating. (Actually, since only one sex is represented in the "strings" under each float of the man-o-war, some proximity may help.) In an even lighter vein, my sailing friends ask whether the port tack individuals yield the right-of-way.

Also floating on the surface are the bunches or rafts of light tan vegetation, the sargassum. If you have a dip net, scoop some into a bucket of sea water. Note the bladders on the blades that keep the weed afloat and the well camouflaged community that lives in the rafts. There's a unique sargassum fish, a seahorse, a gulf weed crab, lace-like colonies of bryozoans on the leaf surfaces, and more.

Overhead you are bound to see the frigate bird soaring on the thermal updrafts. It has the largest wing span, approximately 2½ meters (about 7½ feet), of any bird of its weight. Much is made of the habit of frigates to swoop down on other birds and steal their catches. They also fish actively off the surface of the sea, though they cannot dive or land on the water. Look for the spectacular bright red throat patch of the mating males. This is especially conspicuous when they alight on a high perch.

When not too far from shore you will see brown pelicans and numerous gulls and terns. There's no mistaking the brown pelican with its large "beak that can hold more that its belly can." Most of the gulls and terns are white in the body with black or grey wings. Unlike the gulls, the terns are not as likely to be seen resting on the water and, with their more pointed wings, they seem a bit more agile in the air. Though most of the gulls are laughing gulls, it takes a good birder to pin down the tern species, except for the noddy (brown with a white head) and the sooty (black above, white below). Hopefully you will see the attractive red-billed and white-tailed tropicbirds with forked tail feathers as long as the rest of the bird. Finally you should see what we might call the true oceanographers, birds that are at home at sea and are seldom on land except for nesting. The common representatives in this area are Audubon's

Magnificent Frigatebird

Royal Tern

Brown Booby

Laughing Gull

Brown Pelican

Red-Billed Tropicbird

D.C.K.

shearwater, a medium-sized bird, and the brown booby, about the size of a small goose. They fly and glide quite low over the sea surface.

It is fun to watch the fishing techniques of the birds of the sea. The terns are especially adept as they sight fish from on high, go into a steep dive, and plunge to make the catch. Pelicans and tropicbirds are also good plungers, and boobies not only plunge but pursue their prey underwater. Gulls will often drop shellfish onto rocks (or pavement) to get at the flesh. Not infrequently, however, gulls betray their seagoing style and act as scavengers. Surely you have seen them around landfills.

Most unattractive, but part of the scene to be reckoned with, are the pollutants: plastics, oil slicks, and tar balls that persist from shipwrecks or from tankers pumping the water, used as ballast, from the tanks in which they had carried oil. Unfortunately you may see considerable plastic on shore, and you may experience oil and tar ground into the sand. In the "old days," before plastics, fragments of broken glass were common along the shore. Being well ground along the edges and destined to disintegrate completely (while most plastics do not), the glass did not concern us greatly. In fact, the colored bits of ground glass were sometimes quite attractive.

P.S. Comments on the color of the sea are given in the write-up of "The Surrounding Sea." The winds, the waves and the sea surface conditions are discussed in a separate write-up of "Wind, Waves and the Beaufort Scale." Reading these accounts should also enhance observations while underway.

Coral Reefs

Characteristics

Coral reefs are distributed in a belt around the world in which water temperatures seldom drop below 20°C (68°F). Where the seas within this belt are clear, permitting good light penetration, and where there is a firm substratum in depths generally less than 30 meters, active reef growth may ultimately reach or almost reach the surface.* Such reef development is greatest and most luxurious in waters low in nutrients, thus lacking the abundance of plankton that might diminish light penetration. Furthermore, reefs thrive in relatively high-energy sites, in other words environments with appreciable wave action or strong currents. These requisites are met and reefs are well developed in many locales in the Eastern Caribbean. There are barrier reefs well off the coast, reefs fringing the shoreline, patch reefs scattered in the lagoons between the barrier and the shore, plus variations on these basic types.

Along with the varied and colorful fishes swimming about, the stony corals of the reef create a fascinating scene. There are corals with upright skeletons, vaguely like small shrubs; others are more like boulders; some are vaguely shaped like the human brain; some are thin, almost leaf-like; and there are various

*A reef may occur on a substratum that is currently deeper than 30 meters if it built up when the sea level was lower in the past.

smaller skeletal types. Living as colonies in the carbonate mass they have secreted are the coral animals or polyps, each occupying a tiny hole or other depression. Generally the polyps are expanded at night and contracted, mostly out of sight, during the day.

There is an abundance of one-celled algae, known as zooxanthellae, in the polyp tissues. Their presence facilitates skeletal growth and they are most effective in this regard in daylight when photosynthesis is occurring and carbon dioxide (CO_2) is being used up in the process. In terms of simplified chemistry we recognize that, in the equation involving soluble calcium *bi*carbonate and insoluble calcium carbonate, a decrease in CO_2 will shift the equilibrium toward the latter, in other words toward the essential compound of the coral skeleton. Just how this works in terms of the intricate biochemistry and crystallization involved in corals is not clear.

This pattern of coral polyps harboring zooxanthellae which in turn facilitate coral growth is an impressive example of mutual benefit, or symbiosis, among living things. With rare exceptions, all reef-building corals, appropriately known as "stony" or hermatypic corals, have this specific symbiotic relationship. Without it, as when corals become bleached due to the loss of zooxanthellae, the decreased rate of skeletal growth can have serious environmental consequences as discussed below.

It helps to understand the reef if you think in terms of the coral growth, the consequent breakdown into rubble and fine particles, and the final "cementing" into a firm substratum. To appreciate the breakdown processes you might first take note of the markings where parrotfish and others with suitably adapted beaks have scraped the coral surface for the algae they feed on. Follow a parrotfish around and you may see a cloud of white carbonate fragments expelled from its vent. Next look at some of the smaller sea urchins and see how they lie within little pockets they have carved in the coral. Note, also, how boring sponges are taking a toll. The common sight of feather duster worms peeking out of coral heads hints of what lies within. When you find some dead coral heads (don't sacrifice the live ones), turn them over, break them open, and see what you encounter—more worms, crustaceans, brittle stars, etcetera. You may also see bands of filamentous green algae, some well below the outer surface, yet getting the minimal light they need for photosynthesis. This internal mix of biota adds appreciably to the breakdown of the coral mass.

As noted in the preceding write-ups much of the pulverized material remains unconsolidated and becomes mixed with other fragmented calcium carbonate to form the sand of the region. However, a fair amount becomes consolidated as a firm rocky substratum. This "cementing" is enhanced by encrusting red coralline algae which themselves are embedded with calcium carbonate. These algae, important though they are, may appear as nothing more than a pinkish tinge on dead corals and on the surface of consolidated reef matter.

Though the fishes and the stony corals dominate the seascape of the reef, you will quickly appreciate that the community is teeming with other living creatures. Anemones and gorgonians or soft corals are relatives of the corals but have little or no skeletal elements. In contrast to the corals, they may be expanded and more showy in the daytime. Sea cucumbers and

Elkhorn coral
Great barracuda
Staghorn coral
Mountainous star coral
Pillar coral
Club finger coral
French angel fish
Blue tang
Sand diver
Sea whip
Giant brain coral
Spotted moray
Rainbow parrotfish
Sea fan
Smooth brain coral
Sheet coral
Long Spined Sea Urchin
Fire coral
Star coral
Anemone

sea urchins abound. (Don't step on or let the waves wash you into the long-spined urchins). Some interesting shellfish may be seen; some squid may swim by; and the squid's more sedentary cousin, the octopus, may be seen squirming amongst the corals. Hopefully, even a spiny lobster or two may be noted half-hidden in the crevices. If you have a chance to see the reef at night, you will find that not only the corals but many other forms in this intricate community have come to life. In contrast, some of the fishes seem to have gone to sleep, often hidden or buried. Should you see a sleeping parrotfish you will note that it has secreted and is enveloped in a mucous "cocoon."

The fact that coral reefs flourish in nutrient-poor waters is due to the capacity of such communities to take up nutrients as large volumes of water flow by. In addition, bacteria and the nondescript blue-green algae in the turf fix or transform free nitrogen to a state available for production. In fact reef communities are so proficient in their uptake, nitrogen fixation, and the recycling of such nutrients that the rate of productivity in a well developed coral area is among the highest found in nature. However, since this production is so dependent on turnover or recycling, it does not result in rapid cumulative growth. As a consequence, any yields such as fisheries harvests that are dependent on that growth, are not likely to be large.

Being such highly dynamic systems, coral reef communities are an intriguing mix of fine-tuned ecological needs and considerable resilience. Resilience and the "comeback capability" can be impressive at sites battered by hurricanes. To be sure the older elkhorn corals may have been flattened, but new branches are in the making and the rest of the reef, which a hurricane may have covered temporarily with sediments, often looks as vigorous as ever. On the other hand, reefs can deteriorate quickly when deprived of critical needs such as clear, high salinity water. These conditions are difficult to maintain where human activities result in the nutrient onslaught of pollution, the silt runoff from erosion, and the freshets associated with cleared land.

Yes, man is the reef's greatest enemy. Some think we are being warned of global warming, since in 1987 and 1990 when there were indications that the temperature of the surface water became slightly elevated in the Caribbean, the corals suffered widespread bleaching (loss of zooxanthellae). Having visited the area early in 1988 and 1991 I would say that in both years recovery, at least in some areas, has been impressive.

In the light of the above it is important to reflect on the long range outlook for the reefs. While pollution, silt, freshets, etcetera, tend to be relatively localized, the effects can be cumulative and of widespread concern. Of even greater concern, however, is the possibility that global warming and the consequent rise in sea level could kill off the reefs on a global scale. Current predictions of a sea level rise on the order of 30 centimeters by the middle of the next century amount to about 6 millimeters per year. Though some data show that reef build-up, or accretion to speak technically, may be no more than 1 millimeter a year, rates as high as 10 millimeters have been reported. According to Don Kinsey, one of the most reliable authorities on this subject, calculations of calcification rates reassure us that healthy reefs can grow to match rising sea levels. Moreover, it is conceivable that as sea level rises accretion will accelerate. Finally, even if reefs fail to keep pace, they may subsequently grow to the surface.

One very ominous consideration, however, relates to the above-mentioned loss of zooxanthellae which may result from a rise in temperature. Without the symbiotic role of zooxanthellae, reefs cannot be expected to grow if they survive at all.

Though we know it will be costly, we can cope with the localized pollution and run-off problems. On the other hand the prospects of a global warming trend is a world-wide issue that we may be dealing with "too little and too late." Though the evidence that a warming pattern has caused recent coral bleaching is not conclusive, and recovery from any bleaching that may have been due to warming has been impressive thus far, it would be foolhardy to be at all complacent.

Viewing

The obvious first step in viewing coral reefs is to find good reef areas in accessible locations. Look for high energy settings with good wave action and a good flow of water. Do your observing when the sun is high in the sky and not reflecting off the surface of the water.

While the protected bays and beaches you may prefer for other reasons may not have appreciable reefs, crossing a small island to get to the windward shore or seeking a site "just around the bend" from a protected beach may be rewarding. In fact, by seeking a rocky prominence adjacent to beach settings, you may find good clusters of corals and a choice array of fishes.* Local knowledge always helps in the search for good

*Around jetties and comparable settings you can sometimes find a great display of fishes even if corals are not present.

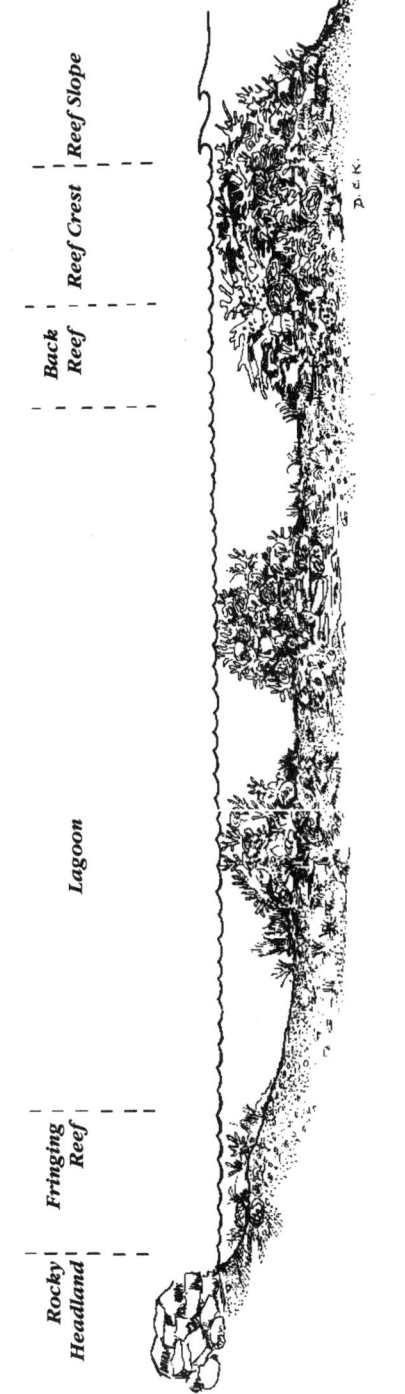

Simplified Transect Across Barrier Reef and Lagoon

An intertidal community occurs low on the headland (see section on Sandy and Rocky Shores).

Fringing reef may be somewhat similar to back reef, plus some sea grasses and algae. Where barrier reef and lagoon are not present, fringing reef may be broader and with some of the coral types of the crest region. Fringing reef development tends to be poor or lacking in front of beaches.

Lagoon varies in width and may be less than 1 to more than 10 feet deep. The floor is calcareous sand with a scattering of small corals (boulder and finger types) and sea grasses. Patch reefs may be fewer than shown, permitting the transit of small vessels behind the reef.

Back reef development may include staghorn, millipore, finger and small boulder corals. Considerable sand and rubble may be present off the back reef slope into the lagoon.

Reef crest may have an impressive development of elkhorn corals with an intermixture of star, brain, pillar and millipore corals.

Reef slope may be characterized by elkhorn corals giving way down slope to more boulder corals and finally mixed coral rubble.

areas. In some locales environmentalist groups are doing a good turn for both the yachtsmen and the corals by providing special moorings.* If these are near enough to the reef to provide good access to the corals, yet far enough away to avoid damage as boats maneuver and drop their anchors, they are a great help.

The mask and snorkel combination is a simple, inexpensive and superb outfit for reef viewing and the moderate skill needed for snorkeling is easily achieved. In comparison, the use of SCUBA (self contained underwater breathing apparatus) is expensive and, except through rentals, is something of a logistic nightmare for the traveler. Also, thorough training is imperative to avoid life-threatening accidents. To be sure, when you use SCUBA you do have the thrill of moving "as one" with the fishes and probing moderate depths. However, much of the luxurious reef development is near the surface. In brief: Viewing with a mask and snorkel can be great! Why not jump in and enjoy the surroundings with this simple equipment while your fellow SCUBA diver is wrestling to put on gear for a dive of limited duration?

Access to a reef depends on the local setting. To reach a typical barrier reef one approaches from shore, often from a boat moving toward the reef and anchoring well behind the crest. Generally the water becomes increasingly shallow on approaching from behind toward the breaking waves that so often characterize the crest and reef-front setting. If you can work the

*Yachtsmen, take heed! Spend the extra time and effort needed to avoid anchoring amongst the corals. My worry is over the damage to corals, but you may very well lose your anchor there as well.

outer edge of the reef front you might experience the most luxuriant coral development; you might see buttresses and grooves reflecting the interplay between the advancing coral growth and the counteracting erosion; and you might find yourself poised at the top of an underwater cliff or drop-off, often known as a "wall." In some locales and in calms this front may be viewed safely, but since the wave action on a reef crest can be quite threatening, judge the situation carefully. If feasible seek local knowledge of each situation beforehand.

In all approaches to the reef bear in mind that currents might be a hazard. And the flow may be particularly dangerous in passes between the reefs for it is obvious that, as water is flowing in over a reef crest, it must be flowing out somewhere else. The buddy system, one of the absolutes of SCUBA, is recommended for all activity in the water. Having an attended boat nearby is a plus. Finally, a shore or ship lookout is highly desirable.

Most snorkelers (and all SCUBA divers) wear swimfins to increase mobility and the ability to cope with currents. Where there are no dangerous currents I often explore without fins and wear reef shoes that come up over the ankles. They enable me to walk over the rubble and to stand when I find a shallow spot, thereby resting and taking stock of my whereabouts.* (You can substitute old tennis shoes for reef shoes but the sand that gets in is mighty uncomfortable.) Getting in and out of the water is also easier without swimfins, particularly from off the beach where my fin-shod companions must walk backwards. (It takes a duck to walk forward nimbly with big web feet.)

*Tread softly. Try to avoid stepping on live corals.

Your mask must fit well and you must be comfortable with the snorkel you select. Call on a friend with experience for help in such outfitting and to orient you to the little tricks (like using a defogging liquid or simply spitting into the mask to avoid fogging). Incidentally, masks with prescription lenses (in broad categories) are available at little extra cost. Hopefully you are using sunscreen but, to avoid having it run under your mask and stinging your eyes, rub it into the skin well in advance and wipe it dry from the forehead and eyebrows before you enter the water. A good quality waterproof sunscreen helps to minimize this problem.

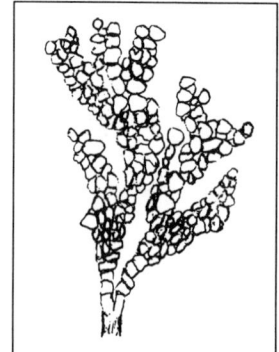

Green calcareous alga, *Halimeda*, enlarged. The plates contribute to beach sands.

Finally and reluctantly I mention glass-bottom boat tours. These cannot duplicate the experience of being submerged in the three-dimensional habitat and some tours take unfair advantage of uninitiated observers with a show of second rate communities and relatively few fish. However, a tour well done can be reasonably rewarding. In the few places where there are submersibles designed for tourists, the viewing can be almost as good as when using SCUBA. In fact, a proper submersible can take you to greater depths than those attainable by duly cautious SCUBA divers.

MOTTO FOR THE REEF VIEWER
Take Only Pictures; Leave Only Bubbles

Atolls in the Caribbean Yes or No?

Charles Darwin unraveled the mystery of atoll formation in the South Pacific and Indian Oceans as he observed these reefs during the *Voyage of the Beagle* (1831–1836). A sequence is involved, as portrayed in the accompanying diagram, and is essentially as follows:

> In the tropics substantial fringing reefs ring the coasts of islands, particularly where the substratum is firm as it is around islands of volcanic origin.
> Such an island may subside with time. This is particularly true through much of the mid-and South Pacific.
> As the island "sinks" the fringing reef becomes a barrier, with a lagoon between the reef and the island.
> Over time the volcanic island may submerge completely with the reef continuing to grow upward as a ring of coral in the shape of the island. Some low calcareous islands may form on the outer reef and various coral knolls and patches may occur in the lagoon.

Anyone sailing or even flying over vast stretches of the mid- and South Pacific is likely to see all stages of such atoll formation. It is as though one were viewing a process in **extremely** slow motion, or by time lapse photography with the lapses on a

36 Atolls in the Caribbean

Young Volcanic Island   **Top Views**

Fringing Reef

Subsiding Island

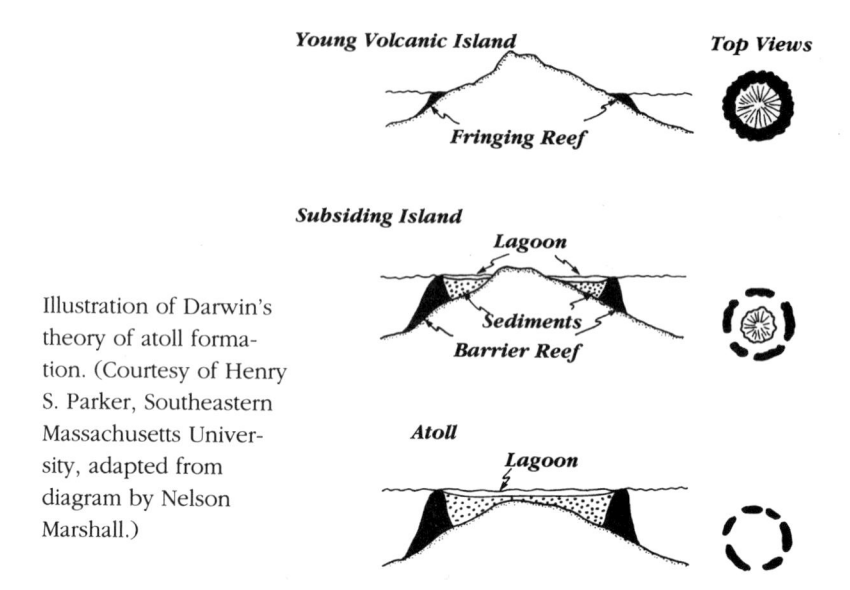

Lagoon

Illustration of Darwin's theory of atoll formation. (Courtesy of Henry S. Parker, Southeastern Massachusetts University, adapted from diagram by Nelson Marshall.)

Sediments
Barrier Reef

Atoll

Lagoon

geological time scale, not just hours or even centuries. From the successive stages seen it is easy for us to piece together this geological history, but think how remarkable it was for Darwin to be the first to interpret it! Had the atoll story been his only contribution to learning, it alone would have established Darwin as one of the outstanding men of science.

And Darwin's interpretation has been confirmed. For example, at Enewetak Atoll in the mid-Pacific, a boring into one of the islands on the rim penetrated over 1,200 meters of calcareous material before reaching basalt. Since the basalt represents the present depth of the "parent" volcanic island, this tells us that coral reef growth, which began as a fringe around the is-

Atolls in the Caribbean 37

land, kept growing upward as its base subsided. While the postglacial rise in sea level has been involved in some of the more recent reef growth and helped shape the upper layers, this could only have a secondary effect since the sea level in glacial times was not much more than 100 meters below the level of today.

Since most of the reefs along the Arc of the Eastern Caribbean are fringing or barriers not far offshore from islands that tend to be uplifting (not subsiding), this is not a region of oceanic atoll formation in the Darwinian sense. However, if we expand our geography and free ourselves from the concept that atolls only form on sunken volcanoes, we do find a few in the Caribbean. Glover's Reef, Lighthouse Reef and the Turneffe Islands complex off Belize illustrate this. They have the basic features of atolls though, rather than forming on volcanic foundations, they have grown upward from other firm submerged features off the coast. This is readily understood when one considers the potential for reef development from a platform beneath the surface and the realization that, where the flow is optimal around the periphery of such upward growth, the flourishing corals may surround a lagoon of less developed patches, knolls, etcetera.

Massive algal-coral ridges that can be more than 300 meters wide are frequently found along the windward reef crests of the atolls in the Pacific. These develop where coral debris has been cemented by and built upon by the coralline algae. Ridge building is not as pronounced along the outer reefs in the Caribbean where a mix of branching, boulder, pillar and other relatively tall corals tend to obscure this process.

Sketch of life on the roots of the red mangrove showing crabs, periwinkles, boat roaches, oysters and a heavy growth of *Bryopsis*, a filamentous green alga, with a snapper and a barracuda lurking among the roots. (Courtesy of Klaus Rützler and Candy Feller, National Museum of Natural History.)

Mangroves and Mangrove Swamps

Some who are insensitive to nature's ways may have little regard for the mangrove swamps that often border the lagoons and mudflats of tropical coasts. It is bad enough if mangroves are cut to make charcoal, to use as poles, or to extract tannins. Of greater concern are the bulkheading and filling of mangrove areas to create dry land for agriculture or development, so called "reclamation." Those who understand nature realize that such uses are an assault on a habitat vital for wildlife, providing a nursery and feeding ground for the young of much of our marine life, enriching coastal waters with nutrients and organic detritus, and helping to offset erosion along the shore. —Does this sound familiar? The need to save the mangroves parallels that of preserving the salt marshes of temperate latitudes and for the same reasons.

Two mangrove species, the red and the black, seldom very tall and forming a wide-spreading canopy, predominate in the swamps of the Eastern Caribbean. The red tends to be more abundant along the lowest, more flooded areas. Its extensive proproot system enables it to pioneer and almost seem to "walk" seaward as sediment builds up underneath. A torpedo shaped young plant develops on the limbs, drops off, sticks vertically in the mud, and continues its growth. The black mangrove tends to thrive on the mud of the slightly higher elevations. Its pencil-shaped, air-breathing rootlets pierce upward through the mud and are very abundant beneath the branches.

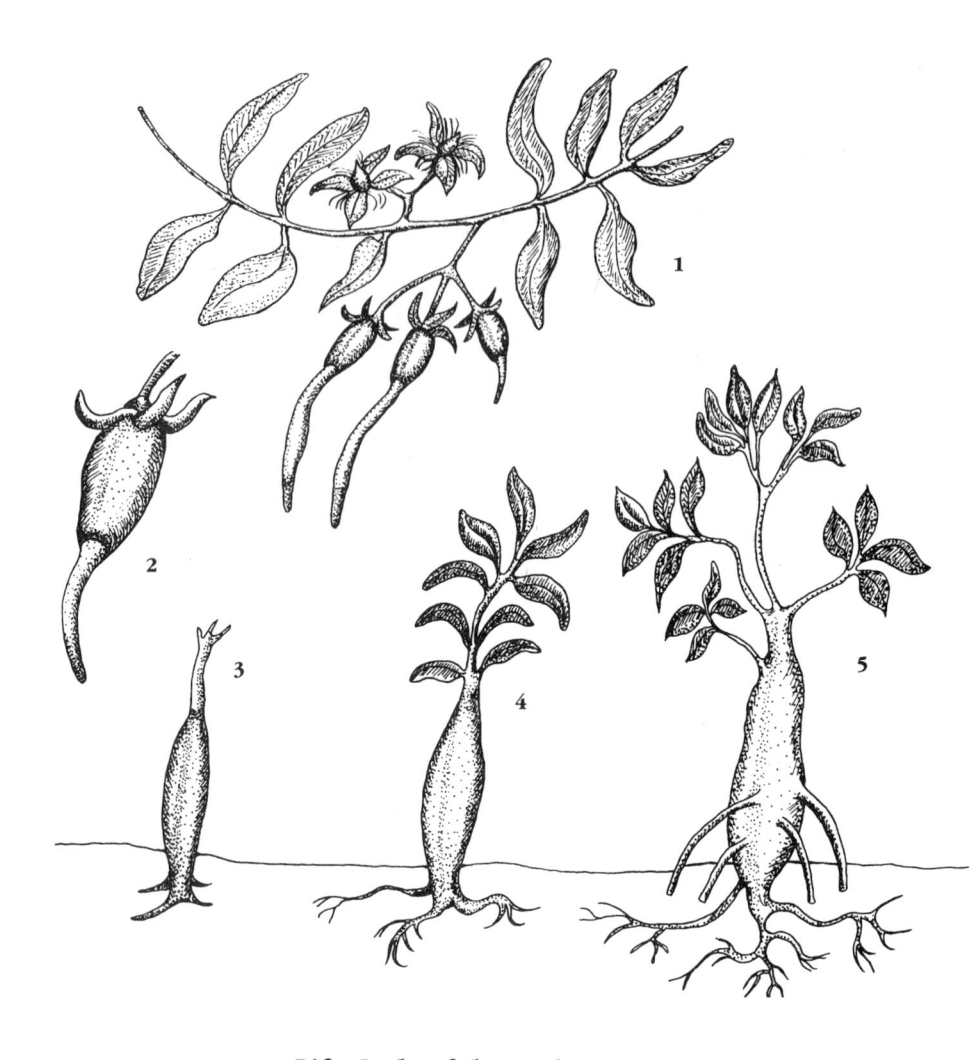

Life Cycle of the Red Mangrove

1. Large dart shaped seedlings or propagules develop from small yellow flowers.

2. Propagules fall from tree.

3. Propagules may penetrate soft mud beneath the tree or may drift, remaining alive for up to one year. Once penetrated into a suitable substrate it begins the new growth as shown in 4 & 5.

In some locales another species, the white mangrove, may be seen scattered among the blacks and extending to still higher ground. Buttonwood may be abundant along the inland border of the swamps.

Anything living in a salty environment such as a mangrove lagoon must master one difficult "metabolic" trick, namely using the water without accumulating the dissolved salt. The red and the black mangroves illustrate contrasting methods, with the former tending to block out the salt as the water is absorbed through the roots while black mangroves excrete salt at the surface of the leaves. Perhaps if other trees and shrubs nearby could do as well, the mangroves would not have the habitat so much to themselves.

Time spent amongst the mangroves is never a waste for a good observer. Note the variety of bird life in the high branches—various egrets, herons, pelicans and sometimes frigate birds, for example. Flights of birds returning to the mangroves to roost at dusk can be spectacular. Large nesting colonies are sometimes found. You may see lizards, insects and spiders scrambling about. Look on the roots and stems, particularly in the region between low and high tide. You will find algae, sponges, oysters, snails, barnacles, sea squirts, anemones, crabs scurrying about and much more. Note the vertical distribution of the life on the roots, particularly from the low tide line to levels above high tide. Finally, look about on the mud and thin layer of water over the floor of the swamp. Numerous small fishes and shrimps thrive on the life so rich in this setting, much of it too small to be seen by the naked eye. This is a nursery area for many important species that are harvested as adults along the coast.

Seagrasses

Naturalist colleagues would rise in anger if I were to fail to discuss the highly productive seagrass beds that are often such a vital link in the shallows between the reef and the shore.

Caribbean seagrass beds are typically a mixed association, with the broad-bladed turtle grass predominating and with various stands of the finer manatee and shoal grasses. Just as whales and dolphins migrated back to the sea during their evolution, these grasses evolved from land plants. And, like the more advanced vegetation ashore, they have true, though very tiny, flowers. Actually, however, they reproduce chiefly from their spreading roots.

Calcium carbonate is commonly seen on the blades of the seagrasses; also, like the eelgrass that is their coastal counterpart to the north, the blades of grass are host to a rich and mixed biota. These leaves, their film of flora and fauna, and the detritus from all of this add appreciably to the food webs of the shallows. Also large quantities of such detritus are being washed out to adjacent areas.

Large and small shellfishes, shrimps, starfishes and their relatives, and much more hide and thrive in the seagrass beds. Even small stony corals, sponges, tunicates and anemones are interspersed there. Many fishes such as surgeonfishes, puffers and snappers are present as juveniles, using the beds as their nurs-

Upper: Blade of seagrass, *Thalassia,* showing calcareous deposits with blue tang in center. (Courtesy C. Lavett Smith, American Museum of Natural History); lower: *Thallasia* bed with spiny lobster. (Courtesy William C. Dennison, Horn Point Environmental Laboratory.)

ery ground. Some parrotfishes, razorfishes, pipefishes, porgies, eels and wrasses, commonly thought of as inhabitants of the reefs, also reside in the grasses.

Grunts, snappers and squirrelfishes are among the hordes that migrate from the reefs to feed on the beds, some as herbivores, some feeding on the resident animal populations. Such migrations are rather spectacular and largely occur during the night. Some sea urchins also move into the surroundings of the reef at night. Due to the intensity of such feeding immediately outside the reef, one often sees "halos" or barren areas around the reef patches. Presumably, if the migrations were in daylight the wanderers would be highly vulnerable to the predatory barracuda, jacks and mackerel that hover above. The daytime presence of the surgeon fishes remind us that not all species wait for nightfall to raid the grass beds.

Rays and marine turtles are among the larger animals that feed on the beds. Rays search for creatures living in the sediment while the green turtle grazes on the grasses. With an appropriate coupling of common names, the grazing manatee of the estuaries is sometimes called the sea cow, as the turtle, manatee and shoal grasses are lumped under the broad heading of manatee grass. Other than local concentrations, manatees are not abundant in the Eastern Caribbean and, where they do occur, they suffer propeller damage from passing motorboats for they swim near the surface and are quite sluggish. There is a myth (a stupid one I would say) that lovesick sailors mistook these homely mammals for mermaids.

In addition to the high productivity of the grass beds and the resulting importance to consumers, their role in trapping and holding sediments is very significant. Without such trapping and stabilization in the otherwise shifting sands, without the accompanying protection of the outer reef, and without the build-up and holding of the shoreline by the mangrove, the shore could be somewhat bold and barren. Of course there are areas where such full protection of the shoreline is lacking, and there are other areas with mixed protection. Such variety is always of interest along the coasts of the tropics.

To conclude: Never think of a seagrass bed as a monotonous underwater field. There are great rewards for the observant snorkeler, some of which can even be seen merely by wading.

The Sandy and Rocky Shore Environment

Some beaches are littered with flotsam ranging from dead sargassum and seagrass, to driftwood and, unfortunately, to plastic trash. On turning over the flotsam you will encounter an array of jumping amphipods known as beach fleas, plus various isopods and insects.

But what about the rest of the beach? Is one to assume that the clear areas or the cleaner beaches are relatively barren? True, this is not a "friendly" environment, being alternately wet and dry, with sand shifting about, and being subject to substantial temperature changes. However, the biota on and in the sands is richer than it might seem. After all, the ghost crabs scurrying about and into their holes must be well fed. In some places herbivorous organisms make tracks across the wet sand as they consume microscopic algae growing at and near the surface. Hermit crabs are often an important part of the team of beach scavengers. On some beaches the small, colorful, wedge-shaped bivalves, known as coquina, that survive on the detritus of the surf may be seen surfing up the swash by the hundreds, then quickly digging into the sand as the water recedes.

As further evidence of life in the compacted wet sand you may see bubbles at the surface or perhaps tiny squirts here and there. By quick digging you may find some of the beach

The Sandy and Rocky Shore Environment   47

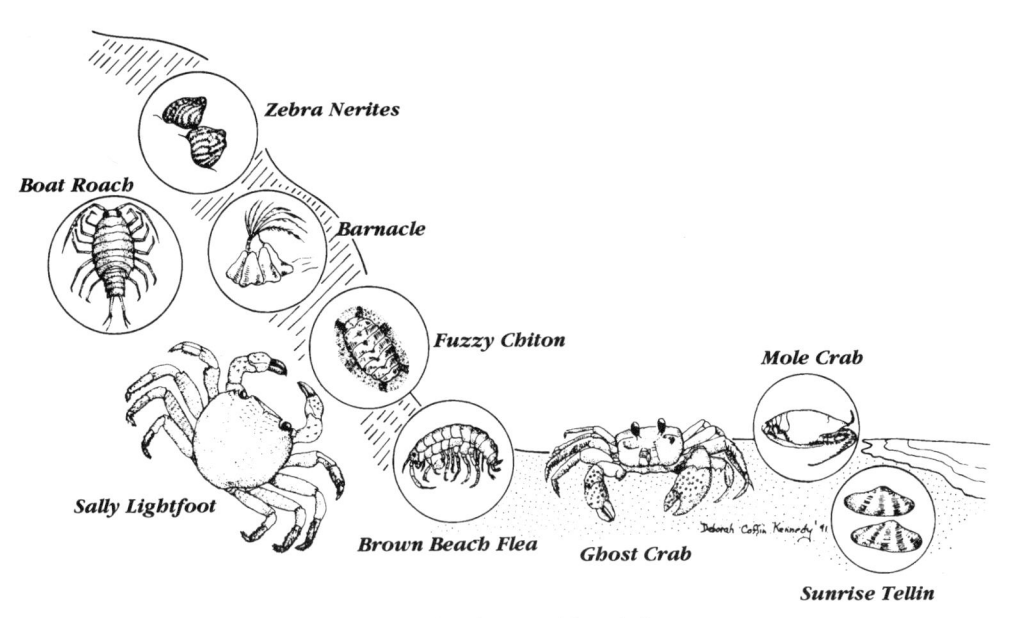

Some Intertidal fauna on the rocky shore and beach front.

fauna within, such as small crabs or bivalve molluscs. Also, tiny holes in the sand may suggest other life, perhaps various worms. Dig to see what's below, but be prepared for frustration as many of the critters are so delicate that they disintegrate when collected and many lie so deep that it takes a lot of shoveling and seining to find anything. Also, some of the holes may be nothing more than the result of drainage from the receding water. Finally, there is a variety of tiny animals, microalgae and organic particles attached to the sand grains or in the minuscule gaps between grains. You cannot appreciate all that is to be found there without special sampling equipment and microscopes.

If you're not fully convinced that the seemingly bare beach is full of goodies, just watch the busy shorebirds meeting their needs with various feeding styles—running and stabbing, pecking, probing, hammering and prying. Watch the sanderlings, for example, that run down the beach in the wake of retreating waves to catch animals revealed in the rolling waters. And marvel, as I do, on seeing the small sandpipers picking from off the surface of the intertidal beach where, try as I may, I see nothing but sand grains.

Mudflats may be found near the beaches, perhaps behind them in some instances. The key to mudflat formation is the lack of currents and waves, thus allowing organic matter to accumulate. Commonly this environment is conducive to mangrove growth and may grade into full mangrove swamp development.

On the rocky intertidal and the beach rock, firmness encourages vertical zonation of attached biota. Varied surface and local conditions, such as the contrast between windward and leeward exposures and the difference between a prominence and a crevice, result in differing vertical zonation patterns, yet certain trends prevail. At the highest level, where there is only occasional spray, the common prickly winkle may be found. The rock may appear to be black high in this spray zone due to the presence of tightly adhering blue-green algae and lichens. Slightly lower, a variety of littorine and nerite grazing snails range about. Around the mid-tide level, barnacles may be common, with limpets and chitons perhaps slightly lower, and with

The Sandy and Rocky Shore Environment

attached marine algae beginning to appear on well-wetted rock surfaces. With the gradation from the very high to the lower levels of the intertidal zone, competition and predation become increasingly important as the variety and abundance of life able to settle increases where the shifts in temperature and exposure become less drastic.

Boat roaches may be seen scampering about high in the intertidal. The lively sally lightfoot crab, with its highly patterned skeleton, is the most spectacular wanderer a bit lower. Try as you may, the chances are you will never catch a sally lightfoot, though you may find its molted exoskeleton. In the submerged environment below low tide there are sponges, hydroids, anemones, tube worms, sea urchins, small corals and a mix of attached algae.

For their food the creatures living on the rocky intertidal zone may, as in the case of many snails, be grazing on microscopic algae growing on the rocks, or they may, as in the case of barnacles, be feeding on the plankton in the water that floods the area with the tides and currents. Some of the snail species and crabs are predators.

Life in the tidepools of the rocky shore somewhat resembles the intertidal and subtidal with variations due in part to predators, such as small fishes and crustaceans that may be confined therein. Differences may also be due to modified environmental conditions as, in the uppermost pools the seawater may be diluted with freshwater run-off and, without frequent replenishment by the tides, the oxygen may be depleted and extreme temperatures may occur. In essence a tidepool may become a distinct environment unto itself.

The Sandy and Rocky Shore Environment

Once again, whether probing the beaches, the rocky intertidal, the submerged regions, or the tidepools, always bear in mind that the underside of anything movable may be the most interesting. But, as a friend of nature, always restore what you have moved.

Additional thoughts as to what to see and how to interpret the shoreline follow in the write-up of "Beachcombing."

Beachcombing

What do the shores of Lake Erie and the beaches of the Caribbean have in common? Very little, except that my eyes were opened to the possibilities of beachcombing when, as a graduate student at the F. T. Stone Laboratory of Ohio State University, I walked the shores of the lake with one of my most influential teachers, Charles F. Walker.

Chances are you will spend many hours strolling the beaches. Here's a chance to fully exercise your powers of observation. Anyone walking along, looking at his or her feet or gazing absentmindedly in the distance, is missing a great deal. The messier the beach, the more there is to see. Most everything washed up has a story to tell.

No doubt you will be on the lookout for shells, but what else will you find? First consider the alignments of the beach wrack. You may see an appreciable accumulation of grasses and flotsam very high on the beach, even in amongst the terrestrial vegetation. This is indicative of the extremes of tide and storm waves. The most pronounced wrack line is likely to be a bit lower on the beach where repeated high tides, including the spring tides of the moon's greatest pull, have built up an accumulation. There may be one or two more lines: one for recent high tides and one nothing more than a faint row of detritus at the water's edge. You are likely to find the most trash, also the more rugged of shells, along the highest lines. For delicate remnants, such as tellins, the tests of sea urchins, and spirula shells, look closer to the water's edge.

Consider the grasses that make up the beach wrack. Are they seagrasses from just offshore or is drifting sargassum a major component?

Presumably you are turning over some of the beach trash and have checked other hiding places finding beach fleas and insects. As suggested in the previous write-up bubbles, tiny squirts and holes in the wetted sand can suggest some life below. Try some digging.

Do you find any of the "sheds" of animals that have abandoned their exoskeletons as they grew? (See explanation of ecdysis in the write-up of "Variety in Nature.")

What do you find in the way of seeds washed ashore? Where did they come from? Are there any remains of upland vegetation littering the shore? This may lead you to thinking about natural dispersal in general, and to contemplating the various ways that plants and animals can populate distant locales.

How much can you learn from a piece of timber washed ashore? Has it been riddled by shipworms? Are there barnacles on it? Any other signs of life? And let's add a bit of mystery. —Is the wood a remnant of a shipwreck? If so, where was the wreck? What do you suppose caused it?

Some jelly-like forms wash ashore semi-intact. Look for the Portuguese-man-o-war, the by-the-wind sailor, and jellyfish. Handle with care—it could be that the stinging cells are still alive, and *very* irritating.

What skeletons have washed ashore? Where did the poor

Beachcombing 53

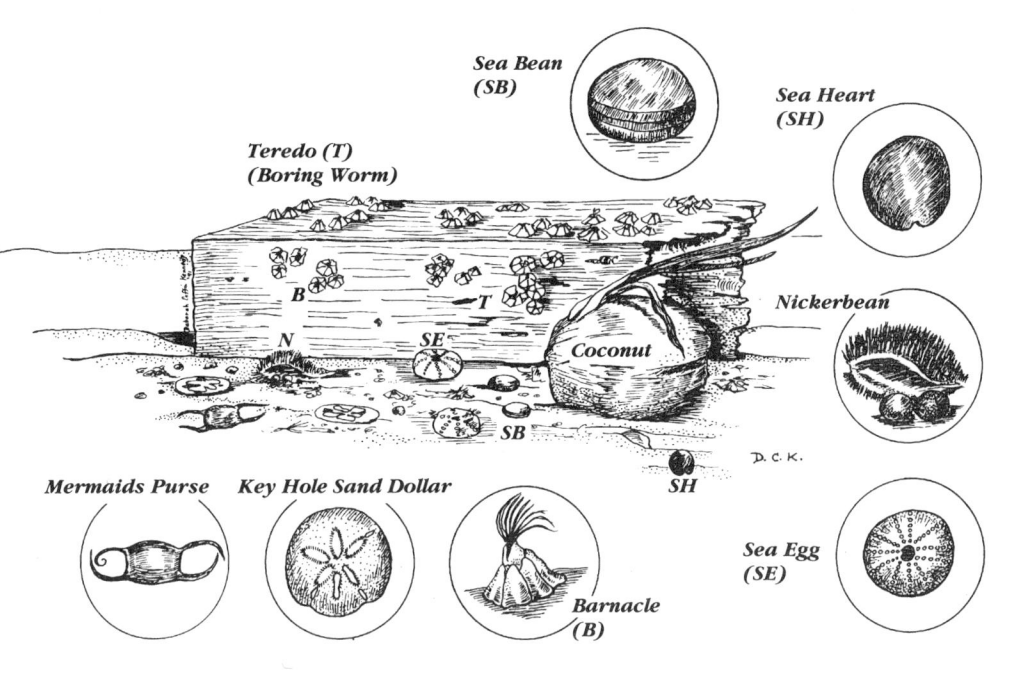

Variety seen when beachcombing.

critters come from? Any thoughts as to the causes of death? Can you tell anything of the creatures' life histories or food habits? —Look at the mouth parts.

A special treat would be to find the tracks of turtles that have journeyed up the beach and back to sea in the ritual of laying their eggs.

Equally unusual are the eggs of birds that nest on the beach. Usually they are well camouflaged and the nests are inconspicuous; however, an angry parent bird will probably warn that you are in an area where you must watch your step. Some nesting birds feign injury as a distraction at some distance from their eggs or young.

All the while you have probably noticed the tracks of ghost crabs running to and from their burrows. Perhaps you have seen the "ghosts" skittering about. If you see other tracks, try to trace them to their origins. What creatures are involved and what are their life habits?

If you pass a rocky headland you can try to interpret the vertical zonation that is often so pronounced. Any tidepools present add still other dimensions to your observations.

One way to pull the story together is to consider the birds you see as you walk along and to ask what they are doing. Are they resting? Are they breeding or nesting? Are they feeding? What kind of feeding is going on? What do their activities tell you about the productivity along the shore and just offshore? (See comments in the write-up of "The Sandy and Rocky Shore Environment.")

Even if you are an amateur you can try your hand at a little on-site oceanography. First, consider whether you are strolling on a high or low energy shorefront, the high being exposed to unchecked wave action, the low being in the lee or behind a good barrier reef. Then consider what the beach sands consist of. Are the origins volcanic? Do you see volcanic rock nearby? Are the origins calcareous? Can you identify any of the pulverized remains? (Here a good hand lens will help.) You may see the blades of halimedas or the tests of one-celled forams.

Is the beach slope steep or gradual? Is there any sorting of the sand particles, for example coarser grains high on the beach left by more vigorous waves and finer sands in the intertidal zone? Do you see small, wave-like ripples resulting from the alternate scouring and deposition of sediments near and below the water line? Contemplating this corduroy effect on a larger scale you may understand why you often experience an undulating pattern of depths when wading in and swimming off a beach. Also, it is common to see lateral undulations, crescent-like as you look along the shore, where waves and currents alternately accumulate sand and wash it away between the cusps. What do the overall beach conditions suggest as to the currents and waves along the shore? Are there any indications that storms have altered the beach lately?

Annoying though they are, the oil and trash on the beach have a story to tell. What offshore shipping routes account for this? Does the trash tell you anything about the currents washing it ashore?

Finally, you must have been checking on the tide. Is it going out or coming in? What is the usual tide range? What are the extremes? Are they largely the result of spring tides, or storms, or various combinations?

So it goes! There's a two-fold challenge here: First, how many of my suggestions can you observe? Then, and more challenging, what can you add from your own beachcombing experiences? All the while, bear in mind that no two beaches are alike. What causes the differences?

Fisheries

You may frequently encounter small scale, artisan fishermen along the coastline. Sometimes they are camping out near reefs and lagoons. These are the "little guys" for whom fishing is a *way of life*. They often go out early in the morning and return the same day. Their gear is simple. Some use traps made of chicken wire; there are various hand-lines, long-lines and trolling rigs; some have beach seines; some have gill nets; and some just dive for conch and lobsters.* Their boats are varied, also simple and modest, and are often the product of their own craftsmanship. The smaller boats are powered by outboard motors; larger boats are powered by inboards. Sails, generally rather crude in material and shape, may serve as a supplement to motor power. The dugout canoes of the past have given way to larger craft.

The artisan fishermen may sell their catches on the beach

*Unfortunately, and in defiance of the law, fishermen occasionally use bleach or other poisons on reefs, leaving swaths of death behind, in fish abandoned to die, poisoned corals, and most everything else that was part of the affected reef area. Dynamiting bits of reef to collect stunned fishes is even worse than poisoning since part of the reef structure is destroyed. Unfortunately you may occasionally see the telltale signs of such outlaw practices.

Though set nets are effective, and not illegal, they also pose problems as turtles and diving birds may become entangled and drown. Even greater losses result from the extremely long drift nets used by large-scale fisheries ventures sometimes operating in the Caribbean. Fortunately, the Organization of Eastern Caribbean States has banned the use of such nets in the area it controls.

or in local markets. Though there may be some middlemen involved, it's often a relatively simple fishing economy. Of the varied catch, groupers and snappers are high on the list of the choice fishes; prior to the present state of depletion, queen conchs were taken in large quantities; spiny lobsters are highly prized and heavily fished.

Two considerations come to the mind when thoughts turn to improving the fishing technology. First, these are fisheries in which artisans enjoy a degree of independence hard to realize in endeavors where major capital or other holdings are essential. Second, the existing simple ways already utilize and often even over-exploit the potential of the resources. Finally, the great diversity of fishes, in kind, in size, and in habitats occupied, is not conducive to commercial operations involving larger vessels, sophisticated gear, and dependent on large stocks of comparatively few species as in the shrimp and tuna fisheries elsewhere.* Furthermore, since the nutrient content of the water is low and is not increased by considerable upwelling from the depths and, since the "fishable" shelf waters are very narrow, the overall potential is not great.

All this notwithstanding, some new approaches to the fisheries show a bit of promise. Coastal fishermen might gear up better to tap the occasional pulses of large pelagic fishes as these may appear off their coasts. Also, without calling for major advances in gear, fishermen are generally encouraged to extend their operations a bit beyond the shelf and the shallow reefs, thus seeking relatively untapped stocks of bigger fishes such as the older snappers and groupers. Attention is also directed to

*The harvests of a distant water commercial fishery, involving large purse seiners, are brought into Mayaguez, Puerto Rico for processing.

Hauling ashore and mending nets (Courtesy of Sandra T. Goodwin of Charleston, S.C.); returning from the fish market. (Courtesy of Craig McLean, N.O.A.A.)

artificial settings which attract fish. There are floating rigs known as FADS (fish attracting devices), sometimes as simple as bundles of palm fronds. Also, by sinking various objects to the bottom, using nothing more than junk cars or heaps of sunken tires at times, man has attempted to increase available fish. The "jury is still out" as to whether such techniques appreciably increase the fish populations or simply attract more fish in areas that are quite featureless except for the coral reefs.

Attracting fish with lights is yet another area for expansion. Squid, jacks and a variety of small but useful pelagic fishes might be gathered in this way. A tried and true technique is to attract the gliding flying fish into a white sheet rigged with a light shining on it and raised from a skiff.

Here and there you encounter some unusual harvests. In some places, for example, there is a fishery for the "sea egg," a sea urchin the eggs of which are considered a choice morsel. Overfishing has necessitated restrictions on sea urchin harvests in Barbados. In Grenada there is a fishery for seamosses, mostly *Gracilaria*, but this is being superseded to some extent by the raft culture of alga (see write-up of "Aquaculture").

The sports fisheries are less frequently encountered throughout the islands. These may grow substantially since they are linked to tourism. We may tend to think of sports fishing in rather glamorous terms, in other words slick boats with dapper captains, trolling with elaborate gear for the big, pelagic oceanic fishes such as tunas, billfishes, kingfish, wahoos and dolphins. The quest is far more diverse, however, and includes hand lining and trolling near the reefs, even diving for conchs and lobsters. Though prohibited in some areas, spearfishing with SCUBA is another sport that is growing in the region.

Offhand, one possible improvement to the fishing economy, namely refrigeration at the landing sites coupled with improved market distribution, would seem especially promising. Fishermen could take advantage of good runs without fear of losses due to spoilage. Also, some of the incidental catch that is now thrown back due to limitations on what can profitably be handled, might be saved. Even so, such improvements might tend to increase the role of the middleman and detract from the artisan nature of the fishery which, in its simplicity, does have appreciable social value.

Whatever is tried to upgrade the fisheries and whatever the hopes and expectations, the possible limits of the resources must be kept in mind. Any new thrust may at first yield larger fishes and larger catches; however, since many of the species caught are characteristically slow-growing and long-lived, such choice catches may soon get fished down. In terms of fisheries theory there is a level at which such populations will continue to provide a "maximum sustainable yield" (MSY), though such a harvest may be less than that of virgin stocks. To realize MSY, rather than over-exploit is, of course, the theoretical management ideal. Applying such theory is difficult, however, where the fish populations are so varied. Thus we are generally forced to rely on and consider management plans based on crude appraisals as to the state of the fish stocks. For example, we can sense that the queen conch is overfished most everywhere and there is a need for more control while the same is not as evident for some of the pelagic fishes. Fortunately the resilience of the tropics (see "Conservation") saves us from the complete collapse of many of these resources.

Some Unique Fish Behavior

How do brightly colored reef fish hide from their predators? The answer is not simple and not altogether clear. Obviously they have access to hiding places where they can hastily retreat and some stay very close to these hideaways until darkness. Some of the dazzling coloring might confuse predators. Against a reef background intricate coloration can blur the outline of a fish. And, of course, distinctive coloration is probably a signal to other members of the same species, as it might serve as an attractant in breeding behavior or, just the opposite, act as a warning in defending territories. Sometimes an extreme in color and pattern is associated with a distasteful or poisonous species or with another species that mimics such an undesirable prey.

There are some impressive cases of camouflage. Many species assume a lighter hue when over the light sand bottom; the peacock flounder is especially noted for its ability to blend into the background. A number of species appear light from below and dark, bluish-gray from above. From below, toward the bright sky, or from above toward ocean depths, such countershading makes it difficult to see these fish. Dense aggregations or schooling may be another primary defense against predators due to the difficulty in "aiming" at individuals. Countershading is common in such aggregations.

Finally, there is the "tropics factor." Bright colors and spectacular patterns are seen everywhere you turn—amongst the birds, the flowers, trees, shellfish, etcetera. Color-wise nature gets a bit extravagant in the tropics and the survival and adaptive aspects, if there are such, are not always apparent.

As you swim about on the reef, look for territorial behavior. Many of the fishes wander within a limited range where they may scare off others, particularly those of the same species. A few may react aggressively to your presence. Many is the time I have been pestered by the nibbling of damsel fish defending their territories. They may be defending gardens of turf algae that they farm on the reef.

In the shallows and about the reefs, be on the lookout for "cleaning stations." This is almost a "believe it or not" phenomenon where large fish literally wait in line at times to avail themselves of the cleaning services of some of the gobies, coral shrimp, juvenile angelfish and other small species which feed on annoying external parasites. Sometimes the tiny cleaners may be seen entering the gill chambers and the mouths of very large fish which gape patiently in return for the service rendered.

A keen observer might sense some of the varied ways fishes communicate. By darting about and waggling its body, a juvenile French angelfish may signal that it is ready to serve as a cleaner, while a parrotfish ready to be cleaned may wait in a heads-up position, and a jack may change to a darkened color on the cleaning station. Elsewhere a barracuda may arch its back like an angry dog. Other fish should take warning and, though it is usually said that barracudas do not harm humans, my very experienced colleague, C. Lavett Smith, tells me he beats a hasty retreat when he sees this posturing.

Clockwise from upper left: trumpetfish, cocoa damselfish, squirrelfish, queen triggerfish. (Courtesy C. Lavett Smith, American Museum of Natural History.)

Some Unique Fish Behavior

Should you see a trumpetfish suspended vertically amongst the gorgonians and other vertical growths, you are witnessing a combined camouflage and feeding adaptation. From this pose it can dart directly down on its prey of small fish and shrimp.

What we might regard as unique sex patterns are not unusual in marine animals. Male jawfish may be seen brooding eggs in their mouths. Male seahorses and pipefish carry fertilized eggs in an abdominal pouch from which they "give birth" to their young. Even more extreme is the sex reversal that is found among parrotfishes, groupers and wrasses. Generally the pattern is for productive females to change completely and become large functional males. Some, being quite aggressive, are referred to as "supermales."

Of course there are as many adaptations as there are species. Watching behavior and considering adaptations is a rewarding art. You may run across remoras and pilot fish with suckers on top of the head enabling them to hitch rides on larger fish and even occasionally on passing boats. The various puffers, porcupine and blowfishes swell by taking in water when alarmed—a pretty good defense against predation. Removed from the water they puff up with air. Frog and batfishes have lures suspended in front of their mouths as attractants to the small prey they consume.

I have only mentioned a few behavior features. I suggest that you look for more in the various guidebooks.

Aquaculture: Present (and Future?)

Aquaculture, or mariculture if you prefer a term focussing on the culture of marine forms, has attracted considerable attention among some entrepreneurs and has been suggested as the way of the future for improving seafood harvests. Actually the culture of fishes in ponds is centuries old in parts of Southeast Asia and there are some outstanding successes elsewhere; however, the rearing of fisheries products is far more complex and more difficult than some advocates suggest. In general, aquaculture is not well developed in the Eastern Caribbean.

One of the most tangible successes reported for the Caribbean islands is the relatively simple cage and pond rearing of introduced European sea bass in Martinique. The induced spawning is relatively simple; little is required for successful grow-out; and there is a ready demand in France for the product. Other finfish are being considered for grow-out in ponds or cages.

Such culture, where the emphasis is on the rearing of easily obtained eggs or young, is simple compared to the efforts to culture the queen conch, the Caribbean king crab, and the spiny lobster. As of a recent report, crab culture is being attempted in Martinique, the Dominican Republic, Antigua, Carriacou, and Grenada. Spiny lobster culture is being undertaken in Antigua and a number of countries are pursuing queen conch culture.*

The capital required and the financial risks are considerable in these more complex ventures. Often the first objective is to develop dependable methods for inducing the adults to spawn. Thereafter, there is the problem of feeding the larvae. Next comes the rearing of the young stages. The final goal is to raise young to adulthood and to have them spawn successfully within a hatchery system. If all this is accomplished reliably and economically, the developer has achieved what is known as successful closed-cycle culture. Since most such intensive culture programs require ample food, provisions to supply this are often complicated and impressive. These often include chambers for growing phytoplankton, much like that occurring in the sea but in greater concentrations. Various vats may be used for concentrations of zooplankton, again resembling food available in nature. A relatively new food potential involves raising turf algae which grow prolifically on screens suspended in flowing water.

Short of a closed cycle, considerable hatchery energy may be directed to bringing the young along far enough to stock them in the natural environment. Since this is the logical first step and is easier to achieve than a fully closed system, such an effort may receive the most attention, particularly in the early stages of aquaculture efforts. Though this may seem reasonable, there are significant drawbacks. First, any hatchery effort may be trivial compared to the reproduction and replenishment of stocks in nature, even when the wild populations seem rela-

*Turks and Caicos, outside the geographic area covered herewith, is a significant center for aquaculture, particularly for the queen conch.

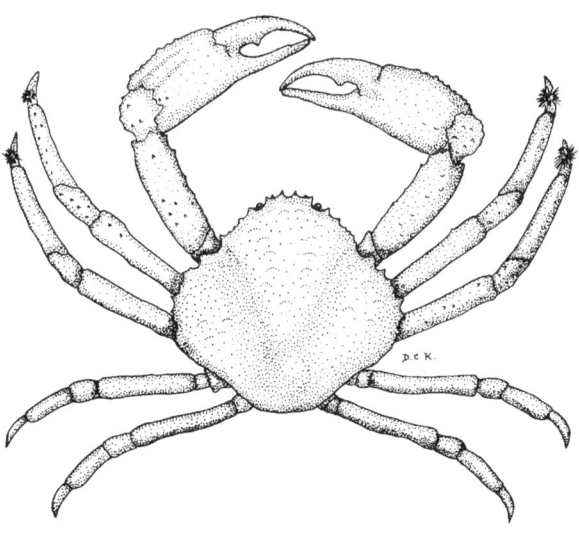

Raft culture of *Gracilaria* on St. Lucia (Courtesty of Allan Smith, Caribbean Natural Resources Institute, St. Lucia); the Caribbean King Crab, a species favored for aquaculture.

tively depleted. Also, restocking can be to no avail if there is little management or control of replenished areas. Then, some skeptics emphasize that young reared in a hatchery, independent of the selection of nature, may lack the desired genetic diversity of the wild stocks. Such comments regarding doubts as to the lack of natural selection do not apply to closed and confined systems where genetic manipulation may be used to advantage.

These considerations are illustrated in the rearing of small queen conch, hoping to replenish these shellfish that are so important to the artisan fisherman but are severely overfished in the shallows. It is now quite practical to collect the egg masses of conchs, rear the larvae in a hatchery, and raise large quantities of small conch. Sowing these in the shallows might seem like a panacea for restocking and is so proposed and practiced by some advocates. But these young that have not survived the rigors of nature may constitute a weaker stock. Realizing that each egg mass produced by a conch in nature contains hundreds of thousands of eggs and that the numbers that may survive can be appreciable even if the percent of survival is very low, it's not at all obvious that production from a hatchery makes a difference. Such is the numbers game so often encountered in hatchery versus natural replenishment.

Three additional mariculture practices come to mind which, in their simplicity, are free of the difficulties of the king crab, spiny lobster and queen conch programs. These involve the mangrove oyster, the West Indian top shell and seamosses. Inasmuch as these require little more, as a first step, than putting out suitable surfaces for the natural sets, such practices are even simpler than the pond and cage culture of finfish mentioned above. On the other hand such grow-out does present problems. Suitable rearing areas must be available in private ownership, through leasing, or in public areas protected by adequate government supervision. The emphasis on the cultivation of seamosses, mostly species of *Gracilaria* and related forms, and less frequently *Eucheuma*, is especially noteworthy and is currently focussed in St. Lucia, Grenada and Carriacou. Since popular drinks and puddings are made using these algae, considerable inter-island trade has resulted. The potential for a large export market exists since seamosses yield agar and carrageenan needed as thickeners and emulsifiers in food and other processed products.

In all of this the reader must bear with the writer inasmuch as aquaculture undertakings change considerably from time to time. Also, one cannot hope to touch upon everything being tried, sometimes in a very transitory manner. In summing up, however, it would be an oversight not to mention closed-cycle programs to raise freshwater prawns as carried out with some success in Puerto Rico, Martinique, Dominica, Guadeloupe and Grenada.

It is difficult to summarize the sponsorship involved in all the above. In a few cases large industrial enterprises are involved, sometimes by companies with subsidiaries that are venturing risk capital on the side. Government backing, as well as support through the assistance programs offered by developed countries, are sometimes involved. Individual enterprise alone may be sufficient where the ventures are simple and do not require large capital investments.

Sea Turtles

Five species of sea turtles are found around the islands of the Eastern Caribbean. The green is the most highly prized for its meat. It has also been taken for calipee, oil and leather goods.* Very abundant in days gone by, greens were readily taken as a choice addition to the food stocks aboard ships exploring the region. The hawksbill turtle has also been in great demand since choice jewelry can be made from its shell.

When a turtle is flipped on its back on deck it remains there, in good shape for a meal when the cook is ready. Since there is a scarcity today, however, the capture of green and hawksbill turtles is generally frowned upon, being restricted and, in some jurisdictions, altogether forbidden. And the same applies to the other sea turtles of this region, namely the leatherback and loggerhead, plus a much smaller species, the olive ridley, which is rare but not unknown in the islands.

All the sea turtles are readily taken if found when they lay their eggs on the island beaches. Once a turtle is turned over, the hunter can go on to capture another at the next laying site. But scarcity and regulations have greatly limited the harvest. In fact scarcity is such that we have only fragmentary, and sometimes questionable, accounts of the nesting locales. We do

*Calipee is a cartilaginous substance of the shell area. Oil from turtles has been used for caulking boats and for medicinal and cosmetic preparations.

Green Turtle

Olive Ridley Turtle

Hawksbill Turtle

Loggerhead Turtle

Leatherback Turtle

know, however, that the nesting behavior is about the same for all species. Females lumber up the beach; turn to face the water; dig a hole with a hind flipper; lay ("drop" may be more descriptive) their leathery covered eggs; fill the nest with sand; and, all within a few hours, lumber back to the sea. During egg laying the females may sigh and shed tears rinsing sand from their eyes. As their flippers have pushed their heavy shells across the sand, the trail left from this journey up and down the beach might remind one of a passing tractor. While much of the egg laying is from June through September, the species and the colonies vary and females may nest several times at brief intervals.

Unfortunately for "turtledom" it is well known that the eggs are mighty delicious, so once they are laid the raiding starts, often led by man but kept up by a variety of four-legged marauders where law and enforcement has held human predation in check. Even so, some eggs make it through and the emerging hatchlings scurry on their bellies, with little legs paddling on the sand, making a "bee-line" to the water's edge. You can imagine the scene that follows with vulnerable hatchlings swimming vigorously through the surf as gulls and carnivorous fishes have a bit of a feast. The plight of the juveniles is greatly eased once they are big enough to be protected by their thicker armor.

Though the adults may remain well below the surface for hours as they sleep, frequent trips to gulp air are essential when active. Some can dive to depths of over 300 meters.

Green turtles graze on seagrasses while hawksbills are turtles of the reefs, feeding to a large extent on sponges. Ridleys and loggerheads have varied carnivorous diets. Leatherbacks, which are somewhat partial to jellyfish, may mistakenly consume plastic bags drifting in the water.—So we have one more argument against plastics.

Marine turtles may migrate far from the locales where they nest, sometimes across thousands of miles of the open sea. Through homing instincts, not fully understood or confirmed, they come back to their birthplace when ready to lay eggs.

So we have an overall story of days past with swarms of turtles in the area. It is a scene that most of us would like to recapture—a prime challenge to conservation practices (see write-up of "Conservation").

The Queen Conch

Walking the shoreline you will come upon large piles of empty shells of the queen conch, *Strombus gigas*. First reaction —"Ah, I'll get a good souvenir." They are there for the taking; however, you will soon note that every shell has a cut high on the spire where the animal attachment has been freed so as to release the meat. Don't be disappointed. After all, curio shops sell these less-than-perfect shells for a fair price.

The shell piles are leftover from an intensive fishery but, if you assume from the size of the piles that the fishery is prospering, you are misled. The shells endure, though the populations are seriously depleted. Queen conchs are particularly scarce in the shallow water closer to shore where they are obviously easiest to harvest, whether by sighting from a skiff or snorkeling. Further, the current use of SCUBA takes an enormous toll both in the shallows and in the somewhat deeper habitats.

Queen conchs grow to 20–25 centimeters (8–10 inches) and the length becomes somewhat fixed when the flared lip begins to develop. After some three to four years, as the flare becomes rounded and thickened, the conch reaches maturity. Copulation takes place during the warmer months, as the male approaches the female from the rear and extends his long, thin penis, called a verge, most of the length of the female to inject sperm. The gelatinous bundle of eggs produced by the female may be extruded at this time; sometimes the eggs are fertilized from previous matings. Each of these bundles contains hundreds of thousands of eggs. Though these egg masses lie on the clear bottom, they are relatively inconspicuous due to adhering sand grains. The females can live as much as forty years and, though they don't grow a great deal, the older conchs produce more eggs.

Once the eggs are hatched as microscopic larvae they drift about for up to thirty days. Though it might seem that they would be carried to distant locales by the currents, Carl Berg and his colleagues, using sophisticated chemical analyses, found that the pin-head sized baby conchs are very abundant near their parent stocks.

Appropriately there is considerable interest in measures that might restore the conch populations, now so seriously depleted over much of the range. The pros and cons of culture efforts are discussed in the write-up of "Aquaculture." But nature may be our best ally. As noted, queen conchs may live for many years; they are not as rare in the slightly deeper, less accessible water; and with some effort egg masses may be collected there. These may be transferred to the shallows and, considering the tremendous numbers in any given mass, such transplants may be our best hope for restoring the queen conch in the shallows. Obviously accompanying management is needed if such undertakings are to succeed.

A live conch is fun to watch. When first handled it retreats within the shell where it is well protected by the horny operculum on the foot. Soon, however, the weird looking orange-ringed eyes protrude at the end of long eye stalks. Then the

Deborah Coffin Kennedy

proboscis may be extended, probing about vaguely like the trunk of an elephant, in the search for food.

If you wait long enough you may see how conchs move. Nothing could seem more clumsy as a conch extends its fleshy foot and lopes along in an apparent random and circuitous path. But there are inactive times as well during which they burrow into the sand, a behavior commonly associated with adversity. Sometimes the burrowing conchs clump, one on top of another.

If we ask about impacts other than the overfishing by man, the story is by no means a dull one. Predators on the young include turtles, snappers, groupers, wrasses, triggerfishes, stingrays and nurse sharks; also tulip snails, spiny lobsters, hermit crabs and octopuses. Conchs are not even completely immune when their shells strengthen since predators on the adults include the octopus with its venomous bite, and the loggerhead turtle with its large strong jaws.

Sometimes a conch will become covered with a profusion of plants and sponges. Some of these are boring types that can penetrate the shell. There is a small brown slipper shell that will attach to the conch's horny operculum. Most interesting of all, however, is the little conchfish that finds shelter under the

The call to the fish market on a conch shell. (Courtesy of Sandra T. Goodwin, Charleston, S.C.)

conch's flared lip during the day, but ventures forth for food during the night.

The queen conch has four companion species in the Eastern Caribbean. The milk conch is almost as big. Progressively smaller are the rooster-tail, the fighting and the hawk-wing conchs. The king helmet, which shares the habitat with the queen conch and is alike in size and comparable with a colorful, massive shell, superficially seems as one with the conchs, but it belongs in a different family. Helmets eat sea urchins and spurn algae, the staple of the conchs.

Queen conch shells are gathered for many different purposes. They are used for stone walls, door stops, garden borders, cameos, grave trim, are ground up to make fine porcelain, etcetera. In contrast to the decorative settings, you are less likely to hear a call to meeting or to market on a conch horn, but such traditions linger on in some of the islands.

The Spiny Lobster

The queen conch and the spiny lobster, *Panulirus argus*. Yes—this *is* the Caribbean! The conch, being so exposed in the shallows, has been severely depleted. The lobster, though of greater value commercially, has better hiding places and ranges to depths a bit beyond those reached by free divers seeking to harvest them. Even so, overfishing is a serious problem.

The spiny lobster is very different from the American, alias Maine lobster. Both have five pairs of legs on what is called the thoracic region, but in the American lobster the front legs, rather than used for walking, are modified as large claws for cracking open shellfish and other food. Nor is the spiny lobster a crayfish (crawfish), though it is sometimes called such. A better use of common names is to apply the term "crayfish" to abundant freshwater species, smaller than either the spiny or American lobsters yet having modified front claws similar to those of the American.

So much for differences. Now for similarities. The various lobsters and crayfish all have large muscular abdominal or tail regions. They can and do move swiftly backward by powerful curling movements of this tail. They move forward and sometimes a bit to the side using their walking legs.

Most of the edible flesh in lobsters and crayfish is the tail muscle; yet in the American lobster there is also considerable flesh in the claws and even some in the walking legs. What about taste? I am sure there are connoisseurs who argue that the American lobster tops them all. I will not contest this except to emphasize that spiny lobster tails *are* mighty delicious. Crayfish, being freshwater forms, have a very different taste that is highly prized in some regions, definitely so in the southern part of the United States.

Lobsters generally reproduce when the water is warm. Male and female lie "face to face" as the male leaves a sticky "tar spot" full of sperm on the underside of the female near where the eggs will be released. When extruded, the bright orange eggs are readily fertilized and become attached to the underside of the tail. With their many thousands of eggs, we now have what we call "berried" females.

The eggs tend to hatch in one to four weeks and the larvae, with thin translucent bodies and long thin legs, drift as plankton near the sea surface for six to twelve months. As they change into juveniles, they settle to the bottom in mangrove, grass bed

Spiny lobster at about 75 feet; diver descending the "wall" of a reef front.

or shallow reef areas. Throughout life, as larvae, juveniles, and adults, growth is by molting, in other words the hard, chitinous outer shell does not increase in size but splits open, the lobster "wriggles" out, and expands a bit before its new shell hardens. The amazing "wriggling out" performance (also mentioned for the sally lightfoot crab in the write-up titled "Variety in Nature") is common for numerous animals that have hard exoskeletons. Spiny lobsters can live some forty years, growing with each molt to a maximum length approaching a meter.

Unique and quite fascinating are the mass fall migrations of spiny lobsters from shallow areas to the edge of oceanic channels. The migrants move forward for several nights and days in parallel, single-file queues with the antennae of each overlapping the abdomen of the lobster ahead. There may be as many as fifty lobsters in a queue, and they may move as much as 30 miles. William Herrnkind of Florida State University, who has been analyzing these migrations, believes that the queues save energy as each lobster eases the flow of water over the one that follows.

There are other similar, though less abundant, lobsters in the Caribbean. Most similar are the spotted and smooth-tailed lobsters. A bit different is the slipper lobster, which is broader, flatter and a bit shovel-like at the forward end. Finally, *Panulirus argus* of the Caribbean is not a loner on the world scene. Numerous relatives are found elsewhere in the tropics and in mid-latitudes. The world traveler may find locally harvested lobster tails on menus far and wide.

Whaling in Bequia

Tiny Bequia, south of St. Vincent, is well known as the whaling island of the Caribbean. The pursuit is for humpback whales, using old-fashioned artisan techniques, in other words sightings largely from shore lookouts, a chase in open longboats, partially under sail and partially by rowing, the take by hand-thrown harpoon, followed by the "Nantucket sleigh ride" which can last an hour or more.

Bequians learned these methods from New England whalers. During the 1860s and 70s, as Yankee whaling reached its peak in the Caribbean, a number from Bequia enlisted aboard the whaleships. After a whaling voyage that ended in Provincetown, Massachusetts, one such apprentice, William T. Wallace, returned to Bequia with a Yankee captain's daughter as his bride, and founded the humpback fishery there in 1875.

The present take may be no more than one whale in a given year; occasionally two or three. Though historically important in the local economy, the practice today simply honors an old tradition in an old-fashion style, with a marginal financial reward and with a lot of excitement thrown in. More power to the Bequians and their modest harvest which may soon fade into history.

Restrictions propounded by the International Whaling Commission (IWC) make an allowance, within limits, for such whal-

Whaling off Bequia. (Courtesy of Nathalie Ward, Woods Hole, Massachusetts.)

ing where it is not highly industrialized and is tied more to subsistence than to commercialization. Specifically, this local fishery is limited to three humpbacks a year, and they must not kill suckling calves or nursing females. Hopefully the limitations projected by the IWC on a global scale will bring a ten-fold increase in the present world population of humpback whales, currently estimated at a little over 10,000.

The humpbacks being chased in Bequia have migrated to their winter mating and calving grounds in the Eastern Caribbean after spending the summer months on the feeding grounds of the northwestern Atlantic. These are the playful leviathans that may almost breach clear of the water as many a whale watcher in the Gulf of Maine area can tell you. Whale watchers in the northeast will never hear the haunting whale sounds, however, for the males that sing do so only during the mating season. Perhaps this relates somehow to courtship.

Whatever the fate of this remnant fishery, the interest in this bit of history and tradition should continue since the Bequia Heritage Foundation is sponsoring a Whaling and Sailing Museum on an original site of the Wallace whaling cooperative.

With whales on our minds it is encouraging to note that the future of humpbacks may be bright, at least in the western reaches of the North Atlantic. Optimism is predicated on sustaining present conservation measures, along with continued educational and interest-generating programs. Especially significant is the protection provided on Silver Bank, a relatively shallow area of some 2,500 square miles centered about 65 miles north of the Dominican Republic. Since this is a major wintering ground where humpback whales mate and bear calves, the action of the Dominican Republic to declare the Bank as a protected sanctuary, is regarded as a major conservation measure. It has been estimated that, at peak season, there may be more than 3,000 humpbacks on Silver Bank.

I am told that a fishery for pilot whales, alias blackfish, has been operating off St. Vincent and perhaps off other islands. Though the IWC has not been overly concerned with such small and more abundant species that generally are not regarded as endangered, increased exploitation could lead to serious declines in their numbers.

Living Light in the Sea

If conditions are right, the surrounding waters may light up in the wake of a vessel, in the crests of breaking waves, or with the splash of an oar. The effect is especially spectacular when fish stir the water and you watch them passing in the glow.* The source of this light, technically bioluminescence, is well known. When present in dense concentrations, certain microscopic plant plankters of the group known as dinoflagellates are a source of the glow we see. One species, *Pyrodinium bahamense*, is the cause of such luminescence in a bay aptly known as Bahía Fosforescente, near La Parguera, Puerto Rico.

Living light is rare among terrestrial and freshwater forms. Fireflies, which can put on quite a show in the tropics, are among the exceptions. But in the sea it is very common. Not only the dinoflagellates but a number of small animal plankters luminesce, and there may be larger, brighter flashes from a variety of organisms, particularly from comb jellyfish. In the darkness of the deep sea, bioluminescence is a continuing part of the scene. In the deep the light does not come from plant plankton, but larger forms are commonly involved, often with highly specialized luminous organs as found in various fishes and crustaceans.

*Where the "heads" on boats are rigged for once-through flushing you are likely to see luminescence in the toilet bowl.

Pyrodinium bahamense greatly enlarged.

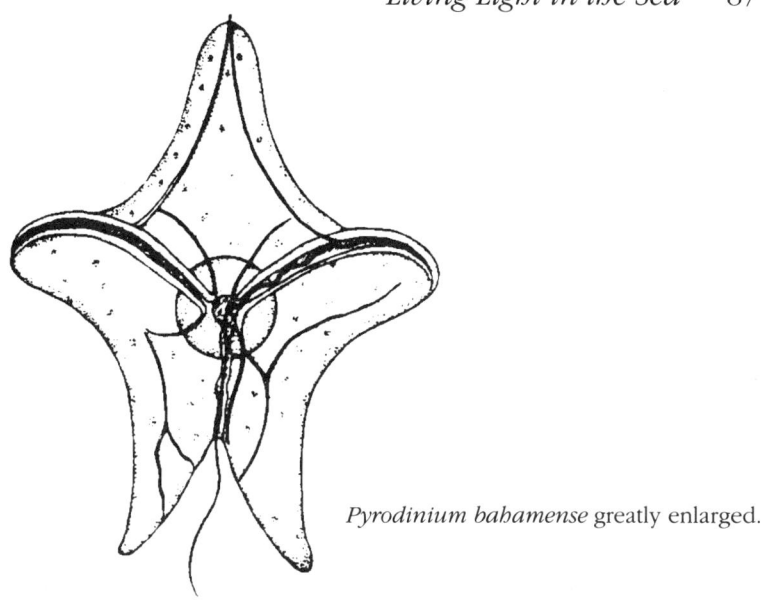

As with the fluorescent lamps in your house, bioluminescence is "cold light" with very little waste heat being given off. Though the dinoflagellates are light producers "in their own right," many organisms derive their light from luminous bacteria. Actually, we don't have a full inventory of which forms produce light and which are dependent on bacteria for their luminescence.

Quite intriguing are questions of adaptation. Various aspects of recognition are often suggested, for example for mating. However, though the energy lost in producing cold light is minimal, one wonders what a one-celled plankter like *Pyrodinium* gains to offset even the minimal cost in energy used.

The Floral Landscape

Tropical and Exotic! —Much of the landscape is characterized by colorful flowering trees,* and by the great variety of flowers and shrubs bearing blossoms both beautiful and ornate. From descriptions alone, who would believe the stunning reality of the passion flower, the African tulip tree, the bird of paradise flower? Or the profusion of ferns, climbing vines and air plants on moist tree trunks? Who would believe the abundance of spectacular wild orchids? *Beauty, variety, luxuriant growth* and *uniqueness* are encountered over and over. The profusion of color in the tropics as manifested by the birds, fishes, corals, shells, etcetera, is even more striking as one looks at the flora.

Beauty is exemplified by trees with large and beautiful flowers as various species blossom throughout much of the year. —There is no barren season. And the smaller flowers, shrubs, vines and air plants are more than a match for this display by the woody vegetation.

Variety in the tropics is exemplified by the thousands of species and diverse forms of orchids, not to mention the hundreds of species of passion flowers and various kinds of ficus trees, including some referred to as banyans, some as strangler figs. The orchid count for the Lesser Antilles is well over 100, plus

*Technically speaking, all trees have flowers for reproduction though, in many cases, they are relatively obscure.

hybrids. It is said that worldwide there may be as many as 30,000 orchid species. There are orchid enthusiasts who seek the wild plants and cultivate them around their homes. Turning to agriculture, there are numerous species, hybrids and varieties of many of the crops. This is well illustrated by the citrus fruits.

Luxuriant growth is generally most impressive in the rain forests and other damp, wooded areas. It is in such settings that some of the unique flora may first gain your attention. There are ferns that stand tall with the trees. There are the leaves of the philodendron which get extremely large as they climb the tree trunks. Bearded banyans are common, with secondary roots hanging from on high and reaching toward the ground. Some trees with hanging roots and climbing stems have literally strangled other trees.* Large buttresses spread above the ground supporting the trunks of some of the trees in the damp forests. Large bamboos, sort of like "big weeds," may abound. There is a profusion of fungi, mosses and lichens; often the tree trunks have a blotched appearance from these growths. Some of the air plants, using trees for attachment only and gaining nutrients and water from the atmosphere, actually have their own "water tanks" or "reservoirs." Examine the "tanks" in the recesses of overlapping leaves of the varied bromelliad group. You may find snails, slugs, insect larvae and even small frogs in these unique "pools."

Warm temperature, humidity, extended daylight and fertile soil, often enhanced with volcanic minerals, contribute to the

*Many species of the genus *Ficus* may act as so-called "strangler figs." The autograph tree, *Clusea rosea*, also strangles other trees, and may be mentioned as a strangler fig though it does not belong to the fig group.

Left: Baobab tree, lower: Gumbo limbo tree in a dry forest. (Courtesy of Gary Ray, St. John, Virgin Islands.)

unusual richness of the flora. Added to this is the adaptiveness and plasticity of most plant life, with shape, size, and at times even color, differing in response to varying conditions. Then man enters the picture, selecting and cultivating varieties and, by cross breeding, accelerating the natural tendency for many species to hybridize. Through the love of beauty and form, the horticulturist has spearheaded the transplanting of species throughout the tropics and subtropics, at first by way of cultivated gardens and nurseries. For most of the flora the spread from such beginnings has been so unlimited that we tend to forget that the *Hibiscus* came from China, *Bougainvillea* from Brazil, that the Angel's trumpet (*Brugmansia* x *Candida*) is the hybrid of two wild species from South America, etcetera. While the Far East is the origin of many species, a fair number came from nearby Central and South American locations. Wild orchids and sea grapes are good examples of plants frequently seen that are truly endemic. Pineapples, which I have seen listed as native by one observer and endemic by another, illustrate how confusing attempts to establish such categories can be. They were present when Columbus arrived but were probably introduced by Indians from South America. Though initial introductions were usually in the cultivated surroundings of homes, hotels, parks, etcetera, many of the introduced forms have become part of the natural setting.

The introductions are by no means restricted to the decorative flora. After his first attempt was aborted by mutiny, Captain Bligh brought the ubiquitous breadfruit from the South Pacific to the West Indies. Bananas, breadfruit, mangos and nutmeg are examples of the numerous crops introduced from Southeast

92 The Floral Landscape

Asia and the East Indies. Nature adds its great dispersal powers as evidenced by the ability of coconuts to germinate after drifting many days, then populating extensive favorable shorelines. The spread has been so effective that the native home of this wanderer remains a mystery.

One can get a sense of the overall impact of the introductions by referring to the appendices where checklists include the "highlights" of the flora. One quickly sees that the overwhelming majority consists of introduced forms referred to as exotics. The "up-side" of this speaks for itself. Sugar cane and other introduced crops served as the backbone of the early prosperity and the flowers and trees are beautiful, or useful, often both. The "down-side" is of considerable concern as introductions, along with man's cultivation and development practices, substantially reduce desirable natural habitat.

Whether introduced or native, it is the spectacular vegetation of the tropics that attracts our attention, yet the more sophisticated botanist will often see flora of equal and sometimes greater ecological importance in obscure and less showy flora. The scholarly botanist may see this in the fields, the swamplands, fence lines and in the wetlands. The variety of the grasses and sedges would be noted. And both botanists and those with commercial interests are bound to call attention to certain trees that may be appreciated more for their usefulness than for show. These include the prolific mahogany trees providing choice lumber, and lignum vitae with wood so dense that it has been used for bearings.

In addition to the rainforests and damp wooded areas discussed to illustrate the luxuriant growth, there is a wide variety

Some plants of the seashore.

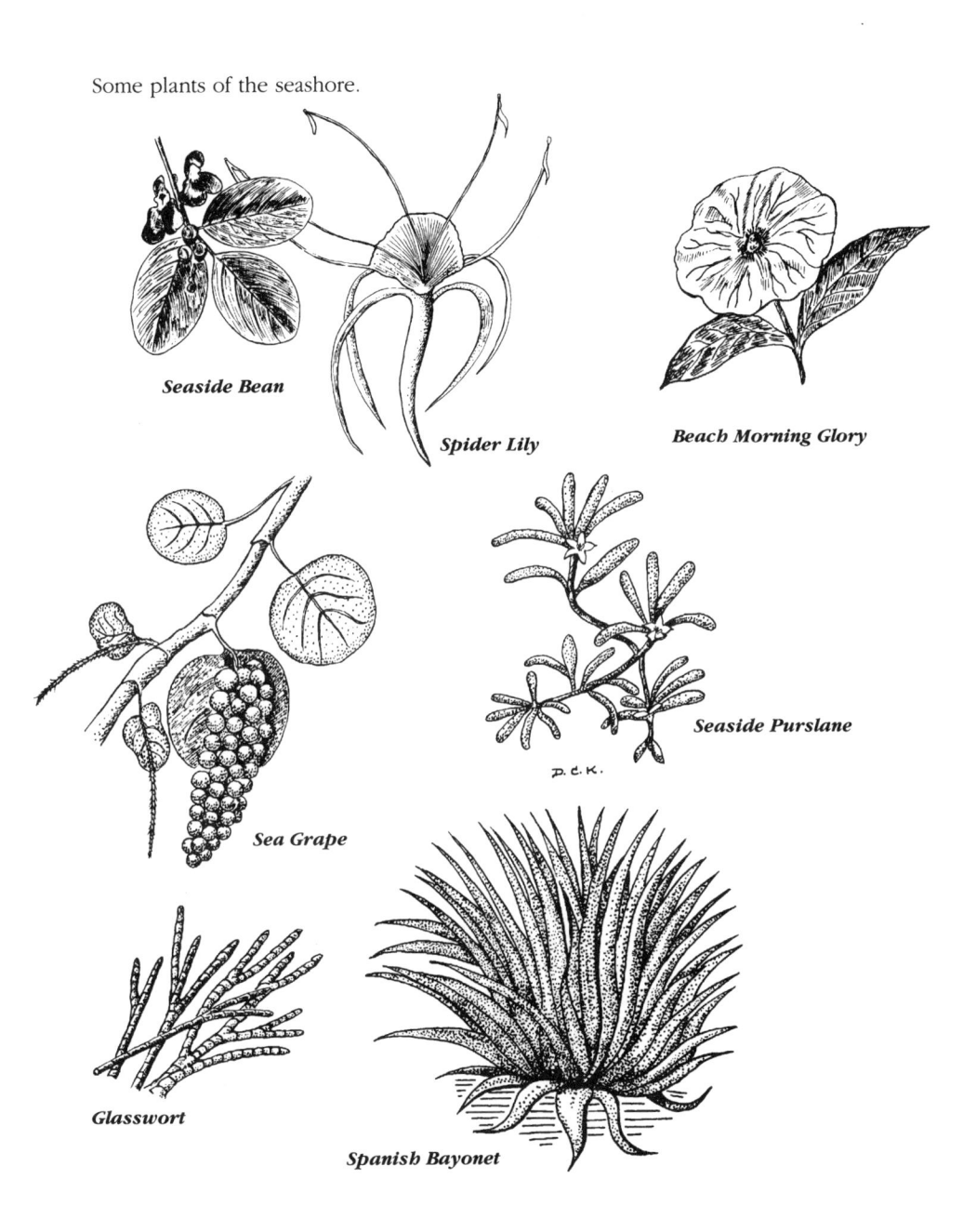

Seaside Bean

Spider Lily

Beach Morning Glory

Sea Grape

Seaside Purslane

Glasswort

Spanish Bayonet

of other natural settings. The transition from one plant association to another may be rather abrupt, as the topography is often rugged and environmental conditions shift rapidly. High elevation and severe wind stress may result in a somewhat dwarfed or "elfin forest" version of damp woods communities. In contrast to the windward slopes, the leeward areas may have extensive dry forest communities. In the driest areas, where an undergrowth of prickly grasses, sticker weeds and brambles prevails, these may be referred to as "thorn forests." Spanish bayonets, century plants and various cacti are common under such conditions. While differing associations of dry forests once covered large portions of the islands, many of these areas have been cleared for cultivation. Inevitably there are some relatively barren areas, some as the result of poor soil fertility or, at best, thin soils that were quickly exhausted. Some areas have suffered from poor agricultural practices.

Along the coastline the mangrove community is the swampland of the tropics, the ecological equivalent of the salt marshes of temperate climates. Here only two or three species dominate and little else has the unique physiological ability to compete. (See section on "Mangroves and Mangrove Swamps.") On the sandy shore, there is a specialized beach flora where low-lying vines and small shrubs with shiny, waxy, salt-resistant leaves predominate. Examples are the seaside bean, the sprawling beach morning glory, the purslane, and the glasswort. Just behind is the coastal realm of the coconuts, which may also stretch to the hillsides. Sea grape and manchineel trees may prosper here. When you are at the shore, whether at a low-lying or a steep site, it is interesting to note that many of the trees

Pineapple. Present when Columbus arrived. May be one of the introductions first brought in by Indians from South America.

have acquired a permanent bend away from the prevailing wind.

In brief, the scene in the Eastern Caribbean is well described by one of my colleagues who, upon returning from a visit said, "The natural surroundings are a special treat, with an abundance of many exotics that one thinks of as unique house plants and as specialties of the flower shops." —How true!

Livestock, Crops, and the Local Market

Much of the livestock and crops has been introduced, some in the early days of exploration. Chickens, cows, sheep, goats, pigs, donkeys, mules and horses are common throughout the region. The cows are often a hybrid mix, commonly including strains of the hump-necked Brahmans. The sheep and the goats look a bit alike since the former lack a thick wool coat and are raised primarily for meat. Just remember, the goats have their tails up, the sheep have their tails down, and you will never be confused. Cats and dogs are almost everywhere.

The crops are a bit more unusual. Almost everywhere there are coconut palms. You will probably be eating lots of coconut. Perhaps you will manage to shuck one from beneath a tree (or have someone do it for you). And the "meat" is used extensively in tropical recipes. Certainly you will see the leaves being used as thatch and for mats, hats and baskets. You may observe that the trunks can serve as pilings. Though commerce and export of the coconut and its by-products are not carried out on a large scale, you may come across sites where the meats are dried for copra, the oil of which is extracted to make soaps and related products.

Bananas and plantains are almost as ubiquitous as coconuts. The bananas are eaten locally and are exported, while the plan-

Avocado

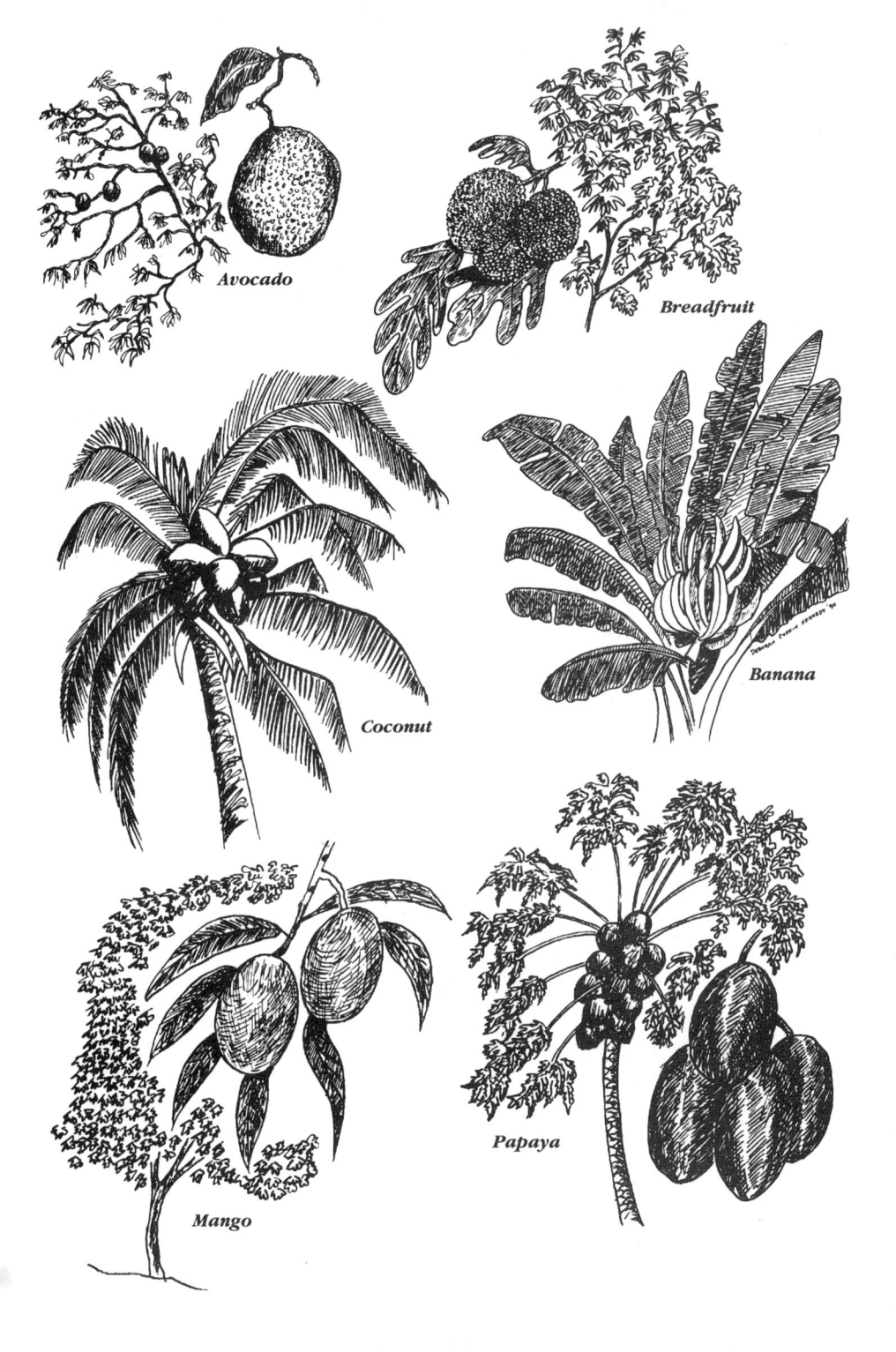

Breadfruit

Coconut

Banana

Mango

Papaya

tain, which is a bit larger, is usually fried for local consumption. The banana plant looks a bit like a tree, but the apparent trunk is basically made up of tightly overlapping leaves. The "tree" produces a single crop in about a year and a half, after which it is cut down to let new shoots grow up from the ground, keeping the cycle going. In well managed plantings the land is allowed to recuperate by letting it lie fallow after a few years. Also, at well managed sites one sees the ripening bunches bundled in blue plastic to keep the product from being riddled by birds and insects if it is to be shipped abroad. Bananas form "fingers up," plantains generally "fingers down."

The harvest of sugar cane was considerably greater in the past than it is today, though it is still common. Today, besides shipping sugar abroad, export dollars are realized where rum is made from the cane. The tobacco harvest was once a major resource, but today it is practically non-existent. Cotton is another product largely of olden times but still exported in small quantities from some of the islands.

Citrus crops include oranges, tangerines, grapefruit, lemons and limes. Again the export is not major, though a fair business may be done in preserves. Cocoa and coffee are produced in some areas but exporting is rather limited. Pineapple is grown very successfully in some areas; it is often sliced and exported as canned goods.

A further listing of abundant fruits would include papaya, avocado, mango, breadfruit and jackfruit. In addition to these more common fruits, there is the christophene, Barbados cherry, guava, custard apple, mammee apple, sweetsop, soursop, cashew nut and pigeon pea. As one listens to natives,

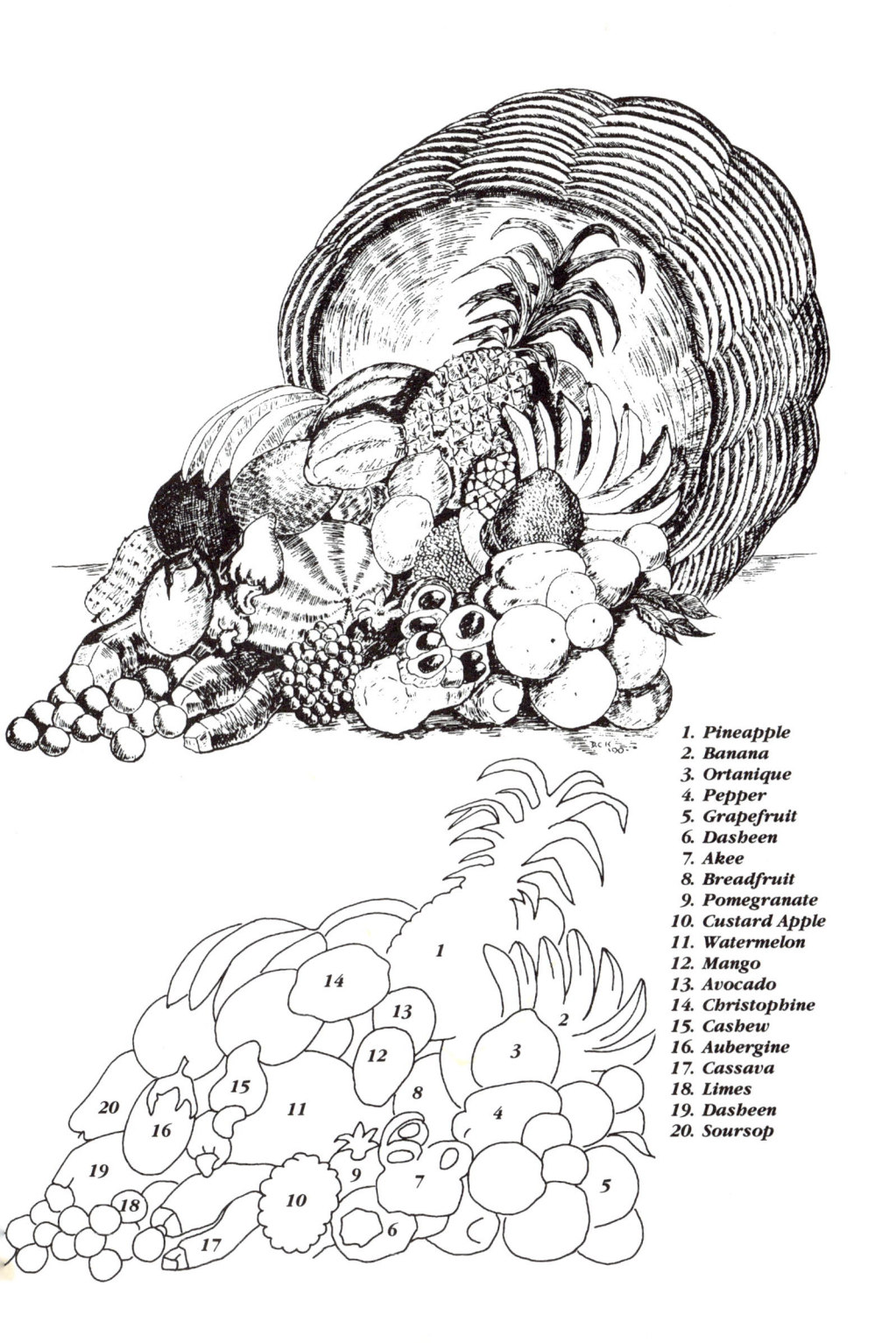

1. *Pineapple*
2. *Banana*
3. *Ortanique*
4. *Pepper*
5. *Grapefruit*
6. *Dasheen*
7. *Akee*
8. *Breadfruit*
9. *Pomegranate*
10. *Custard Apple*
11. *Watermelon*
12. *Mango*
13. *Avocado*
14. *Christophine*
15. *Cashew*
16. *Aubergine*
17. *Cassava*
18. *Limes*
19. *Dasheen*
20. *Soursop*

seemingly countless reports unfold as to unique uses of fruits and leaves for juices, soups, flavoring, etcetera. For example, the nut of the tamarind tree and the fruit of certain passion flowers yield juices used for tasty drinks.

Garden crops that abound in the market run the gamut of the well known: carrots, beans, peas, tomatoes, beets, cucumbers, turnips, sweet potatoes, squash, spinach, radishes, onions, peppers, corn or maize, eggplant, peanuts, pumpkins, melons and watermelons, cabbage, okra, lettuce, parsley, and so on. In essence it's a very familiar list to tourists but we can add the tropical touch with examples such as arrowroot, akee, carilla, cassava, yams and dasheen (alias taro or edoe).

The people of Grenada would feel slighted if we failed to note the pungent and flavorful spices for which their island is so famous. The most familiar are nutmeg, also mace made from the red covering of the nutmeg seed. Other spices include black pepper, ginger, Tonka bean, vanilla, cinnamon, clove and allspice.

Variety in Nature

As various natural features have come to mind I have written about them in the foregoing brief accounts. Still there are a number of points left over which, for convenience, are assembled in this section. You may have your own points to add, from your observations and from other sources such as the comments in various guide books. Accordingly space at the end of this section and some blank pages at the end of the book are provided for *your* notes.

The shed skeleton of the beautiful sally lightfoot crab, that you are likely to find if you explore the rocky shore, is a reminder of the fundamentals of growth of all crustaceans.—They must shed the outer skeleton if they are to increase in size. This ecdysis is amazing. Just to look at the intricacies of a shed exoskeleton and to realize that a creature neatly extricated itself, even from the finest recesses, joints and branches, is to marvel at the process. If you are lucky you may actually witness a shedding.

Many organisms have remarkable powers to regenerate lost parts. Some can even grow whole animals from nothing more than bits and pieces. Pieces of sponge and parts of starfish are notorious for this ability. Possibly you will see a starfish consisting of one arm of normal size and four little arms growing out to form a new whole. If so, you can figure that somewhere else there could be four other full-sized arms similarly undergoing repair.—Lesson to be learned: Don't figure on controlling a predatory starfish population by chopping the animals into pieces. You only create more starfish.

On a lesser scale animals may use regenerative powers for defense. Should you grab a crab by the claw, chances are it will "gladly give you" the appendage, that is if the crab has not nabbed you first! For the crab it's a loss easily repaired. You may see some crabs with undersized claws since regeneration does take a little time.

Tear open a small piece of sponge, using gloves for protection against the spicules present in many species. You will see that it is made up of countless canals. Whips beating from the cells in the sponge draw water through these canals, generally in at the sides and out at the top. The canals of the sponge are havens for critters who live within. You should find more than a dozen species and ten times as many individuals. When through, return the pieces of sponge to the water. —They may regenerate.

As you snorkel or SCUBA across an open lagoon, you may see countless mounds of sediment. This is the work of callianassids, a type of shrimp which feeds on the organic matter adhering to sediment drawn into chambers that extend to a depth approaching a meter and a half. The callianassids expel the "used" calcareous particles as volcano-shaped cones overlying their burrows.

Believe it or not, the ubiquitous barnacles are closely related to shrimp and crabs. If you were to see their drifting larval stages, you would recognize the relationship. As adults, however, they settle on their backs, surround themselves with calcareous plates, and kick their feet upward to catch passing food particles. Look carefully at submerged barnacles and you will see the feeding action.

Anyone maintaining a wooden boat has a never-ending battle with teredos, alias shipworms. From their elongate borings, which you may see in driftwood for example, you might think they were worms, but they are not. Two small shells at the drilling end of their wormlike bodies tell us they are bivalve molluscs. In other words, they are related to clams but their valves are adapted to boring.

Sometimes, when all else is quiet as you lie at anchor or alongside a dock, you may hear a fizzing sound as if you were holding your ear close to a newly opened bottle of soda. Unless you are a veteran to such an effect, you may launch a search fearing trouble somewhere in the hull of the boat. Actually, it's just a chorus of snapping shrimp on the bottom beneath the hull, each with an enlarged claw adapted to produce this noise, presumably to defend its territory and even to stun small fishes for food.

Certain small shrimp make their homes in the crowns of anemones, immune to the hosts' stinging tentacles. From this well-protected haven they venture forth for food, including the parasites they pick up on their missions to the fish-cleaning stations (see write-up under "Unique Fish Behavior").

The stinging cells, technically nematocysts, that cause the severe sting of the Portuguese man-o-war, are present in all jellyfish. Though they are also present in all hard and soft corals, only the fire corals affect humans. These stingers are neat little units of living artillery. A poisonous thread with barbs on the end is tightly coiled within each cell. Contact with an external trigger at the surface of the cell causes the barbed coil to discharge and the target is hit.

Where the long-spined urchins are common you may notice that they are not randomly distributed. For reasons not altogether clear, they gather in clusters or, as one of my farmer friends put it, in "herds."

If you collect an Agassiz' sea cucumber, keep it in a bucket and change the water frequently. You may see its "resident" pearlfish. The fish escapes at night to feed and returns every morning to its home inside the sea cucumber, "worming" its way in backwards through the cucumber's anus.

Smoke screens? Yes! An octopus or squid may shoot out an impressive cloud of black "ink" as it moves swiftly away when challenged or frightened.

Nudibranchs are snails without a shell (nude) and with elaborate gills (branchs) on their backs. Some, with their colored flesh and showy gills, are among the most beautiful animals to be found in the shallows. To search for these beauties and their close relatives, sea hares and sea slugs, look on and among the sea grasses, on sea fans, on mangrove roots, and so forth.

Many of the egg cases seen along the shore are quite distinctive. There are the collars of mixed sand and gelatinous secretions filled with the eggs of the moon snail. The very distinctive mermaid's purse is rectangular, shiny black and pillow-like, with a curved tendril at each corner. Each purse has (or had) a miniature developing skate inside.

Variety in Nature 105

Walking on the beach one is also likely to encounter distinctive plant seeds with resistant outer coats and air-filled spaces for buoyancy. These include the yo-yo shaped sea bean and the smaller sea heart, both being seeds of climbing vines. The familiar coconut is another seed that may have washed up from afar.

The coral jewelry sold by native fishermen and in the shops, is from black coral, usually taken from depths greater than 30 meters. The art of making jewelry from this coral has not advanced to the extent that it has in Hawaii where pink and gold as well as black precious coral is obtained. While some of the polished pieces and decorative shapes do make fine souvenir jewelry, the promotion of this market can lead to serious depletion of threatened coral stocks.

You may notice that the inner coating of mollusc shells tends to have a bright and colorful lustre. Since this surface is laid down by a fleshy mantle you might surmise that an irritant, say a tiny grain of sand, might cause the mantle to lay down the same lustre around it. Could this become a pearl? It's possible, though many such secretions just form as lumps on the inside of the shell. Pearl formation is rare, estimated as one in every ten thousand for conch shells. They bring a high price, but are not precious gems since, unlike the pearls of Japanese oysters, their rich color and lustre breaks down over time.

Trichodesmium, a planktonic blue-green alga, defies my generalization that you will not see planktonic forms with the naked eye, certainly not plant plankton. It's not that *Trichodesmium* is large but, with many cells clumped together, the effect is an abundance of small reddish brown dots in the clear blue water. In fact, where it is sufficiently abundant as is common in the Red Sea, the water may appear red. Ecologically this alga, which may amount to as much as three-quarters of the phytoplankton in these waters, is very important. It is not only highly productive but, by fixing free nitrogen, it adds nutrients to the surrounding waters.

The sea skater, *Halobates,* is an insect of the water strider group that literally skates on the ocean surface. It is limited to tropical seas and has the unique ability to complete its life cycle at sea, quite in contrast to oceanic birds all of which must return to land to lay their eggs. If the water is extremely calm you might look for the skaters, though they are very small. That they have also adapted somehow to surviving on the surface in rough water is quite amazing.

On being artificially fed creatures may lose some of their ability to fend for themselves in the wild; also the food offered may lack certain critical nutritional requirements. Nevertheless, one may want to cast a few bread crumbs or crackers about underwater to attract fishes. It often works and, on a modest scale, is probably as harmless as tossing crumbs to gulls. Of course, it's fun to watch a trained guide feeding such fish as morays by hand (even mouth to mouth), but leave this to those with experience in such showmanship. The behavior of the creature being fed may not be what you expected and could even be dangerous.

Due warning is given (see write-up titled "Beware") that eating the flesh of puffers can be dangerous. It is known that, in the voodoo cult in Haiti, puffer flesh is added to the mix in powders used by bokors in "zombification." There's a running controversy as to whether the fish's poison, tetrodotoxin, induces the zombie effect. And there are questions as to how the bokors work and what the zombie state, so-called "living death," actually is.

Over and over you will hear of the dangers of the fruit and sap of the manchineel tree that is so common along the coasts. Columbus and his men learned of this early on as they ate the tempting apple-like fruit and suffered great pain, reportedly going temporarily mad.

Many of you have probably kept bryophyllums (also referred to in the checklists as the leaf of life) in your homes. As new plants grow out of the creases of the leaf margins, they are fun to watch. You can see the same new growth in the wild in the tropics.

Occasionally you may see a mongoose scurrying across the road. Introduced for rodent control by the owners of sugar plantations, they have held down the numbers of the Norway rat, but not the black rat. They are great snake killers; in fact, I have heard that the paucity of snakes in Grenada is due to an abundance of this predator. As with many species that are not native, however, the net effect of this introduction takes a tremendous toll on the natural balance. The mongoose preys on a wide variety of ground-living forms and raids the eggs in bird nests.

Other mammals, reportedly introduced but found in the wild and not at all rare in given localities, are the agoutis, armadillos, opossums and raccoons. While these may be hunted in some locales, the more spectacular game includes the fallow deer introduced in Barbuda. Escaped monkeys have established populations that are sometimes rather large, even a nuisance on various islands.

Pigs were introduced by Columbus and found the region so much to their liking that they prospered both domestically and in the wild. I have it on good authority that the wild pigs make for good hunting in some locales and that there are even restricted hunting seasons. Thus, in my imagination, I have the seemingly ridiculous picture of a big, lean and mean pig charging a brave hunter who, much like those in African movies, fires the lethal shot "in the nick of time."

With an eye to the sky at dusk, you should see some of the various bats that abound. There are insect-eaters, fruit-eaters, some that imbibe from succulent flowers and there are even fish-eaters.

Hummingbirds make the most of the Caribbean. You will do well to see four of the seven or so species to be expected in the area. They are all extremely beautiful and very small. In fact, the male bee hummingbird which occurs in Cuba is the smallest of all birds.

Tame lizards scurrying about the dwellings help hold down the insect population. Enjoy them. Many is the time I have looked for a bird indoors only to find one of the little geckos singing away, perhaps placidly holding on upside down with its suction cup toes. These little guys have the unique ability to shed their tails when attacked. Presumably, the idea is that the pursuer will eat the tail while the gecko runs free to grow another. Prominent among the lizards are the abundant chame-

Upper: sand mounds of callianassid, alias volcano shrimp; middle: the handsome face of an iguana; lower: banded coral shrimp, a cleaner species. (Two upper photos courtesy of William C. Dennison)

leons which, as though bragging but in reality expressing anger, inflate their colorful throat pouches. Some have the ability to change color almost instantaneously.

At night, and when time and conditions are right, the loud choruses of tiny frogs singing in the trees and wetlands will add to the fascination of the tropics. Don't be surprised if a big frog croaks so as not to be overlooked. One species was introduced to help control pests in the cane fields. A very large frog, with legs so tasty that it is known as "mountain chicken," is now considered endangered, and I am told it is found primarily in Dominica and Mountserrat.

Though iguanas, which look superficially like miniatures from dinosaur times, are not widely encountered, you may see them in great numbers at some sites. Some of the islanders relish iguana meat.

Hopefully you are not a tenderfoot with a great fear of snakes. St. Lucia boasts of boa constrictors which can be up to ten feet long, so I have heard. They strangle smaller prey, not humans. Harmless black snakes can be quite large as well. To be sure the fer de lance, found in both St. Lucia and Martinique, has a very poisonous bite (see write-up titled "Beware") but this snake is becoming quite rare.

Though not widely distributed, tortoises, which resemble the gopher tortoise of the southern United States, do occur. On the ground you may see the open entrances into their cave-like burrows.

A special treat is in store should you find the beautiful striped tree snails in the damp forest but the odds are not in your favor. Since they are both attractive and readily plucked from the tree trunks, a heavy toll has been taken by collectors.

Variety in Nature

Probably to most people the big creepy land crabs crawling over the ground and up the trees are a bit bizarre—almost like something from outer space. Actually they are harmless, even good to eat. You tend to find them in low areas rather near the water, sometimes making the large burrows they crawl into to keep their gills and eggs moist. Around resorts you may see signs: "Crabs Poisoned." Proprietors, who are killing the crabs to keep their guests happy, are warning the locals not to eat the poisoned flesh.

Though the tropics may be thought of as the domain of large and beautiful butterflies and though these and other insects have been neglected in these natural history notes, I feel compelled to at least mention the monarch. This is a cosmopolitan butterfly of North and South America, Australia and the East Indies. You may know it "back home." The monarchs of the temperate regions migrate thousands of miles to their wintering areas. Actually, the Eastern Caribbean populations do not stray far; yet true to the species they are strong fliers.

Termites must think they "own" the tropics, at least as far as anything made of wood is concerned. They look like white ants but they are not ants—they don't have the pinched waist. The key to their success is their intestinal fauna of one-celled forms that can digest the cellulose of wood. You may see massive dark brown termite nests or mounds on tree trunks and plant stems. In nature the destructive role of termites is very important. Without these and other forms of life that cause decay, forests would be littered and cluttered with fallen trees.

Mahogany butterflies—cockroaches that is. You deserve an A+ if you can operate in the Caribbean without at least a minor infestation in your surroundings. Of course no one intentionally brings cockroaches on board or into the house, but the eggs sneak through in packing materials and in the crevices of cartons. You may be forced to use strong poisons to combat roaches, which is a problem if children and pets are with you. In the extreme, fumigation may be necessary. My suggestion: Don't get too worked up; tolerate a few. You will probably end up doing that anyway.

The metal bands on the trunks of coconut trees guard against another menace, namely, rats that climb the trees to destroy the fruit. On some highly manicured lawns you may see various trees painted white to about head-height. Reportedly this is just a fad, supposedly decorative.

Prostoglandin, a mammalian hormone important in the control of blood pressure and metabolism, the aggregation of blood platelets, stimulation of smooth muscle, and active against leukemia, occurs in large quantities in the black sea rod, *Plexaura homomalla*. This is one of the gorgonians growing on the bottom amongst the corals. The history of this potential resource typifies the finding of useful drugs in nature. The discovery aroused considerable interest; the use seemed most exciting; plans for large-scale harvesting were projected; then, as chemists went to work, they found they could synthesize the hormone more readily than harvest it. —This was all to the good in terms of ecological consequences.

In the continuing search for useful drugs and other extracts much of the focus is on the tropics since, due to the density of competing populations, many species have unique chemical defenses. In addition, folk medicine and other native practices may lead to new possibilities. The fact that a natural extract may

Variety in Nature 113

ultimately be replaced by a synthetic compound does not negate the value of the initial search that led the way.

So much for additional odds and ends that come to mind when contemplating the natural features of the Eastern Caribbean. Now it's *your* turn.

Variety in Nature—Readers' Notes (also use blank pages at the end of the book.)

Winds, Waves, and the Beaufort Scale

The tradewinds that prevail in the Caribbean are basically very friendly to anyone traveling by sea, particularly when sailing north and south, or to the west. And they help make things quite comfortable ashore. You can enjoy steady, fresh winds (on the order of 15–25 miles per hour) and all blowing from one general direction, the northeast to east quadrant. The wind-driven waves, building up across an unbroken stretch of thousands of miles to the east and thus the product of an enormous fetch, may be quite large, particularly to the east of the Island Arc. Being long from crest to crest, however, they are comfortable for sailing. Hundreds of boats, generally taking off from the Canaries following the lead of Columbus, safely cross the Atlantic in these latitudes. Hundreds also sail the northeast and southeast trades enroute around the world via the "coconut milk run" across the Pacific and Indian Oceans.

Caribbean sailing conditions are not always so ideal. Sometimes the winds fall off at night, and there can be spells of boring calm under a blazing sun. At the other extreme are occasional squalls and perhaps a few days of peak wind velocities on the order of 35 miles per hour. With such high winds, with waves that may be more than four meters or 15 feet from trough to crest, and with white foam blowing in streaks, no sailor is going to feel altogether complacent. Also menacing are the high winds "funneling" between the islands and the waves piling up into a confused state in the island passes.

Actually, confused wave patterns are more the rule than the exception and the most common contributing factor is a swell from a different direction (say from a storm in the North Atlantic) interacting with the regular train of waves built up by the tradewind. One pattern building on another can create an occasional monster wave. There is an old saying that every seventh wave is a big one, but the timing is not really that consistent. In the extreme, the rarest of these monsters are called "rogue" waves—a very appropriate adjective.

Obviously none of these remarks regarding the tradewinds apply when a tropical storm or hurricane is imminent, or when a "norther" blows in from a weather front of the temperate latitudes.

One might think that my remarks as to the good sailing in the trades contradict earlier suggestions that, while traversing the Caribbean, it is most enjoyable to follow a route in the lee of the Island Arc. There are two points to consider in this regard. First, why not enjoy the leeward sailing when it can be so comfortable? Second, if you are windward of the islands you are taking unnecessary risks, especially when relatively close to shore rather than far out at sea. To explain this it is useful to discuss the basic dynamics of wave motion.

In contrast to the progress of a train of waves moving across the deep ocean, the water within the waves rotates in vertical circles twice the depth of the waves. Such water has only a slight forward motion from the interacting forces of gravity and inertia. You can sense this rotation by watching a chip floating

116   Winds, Waves and the Beaufort Scale

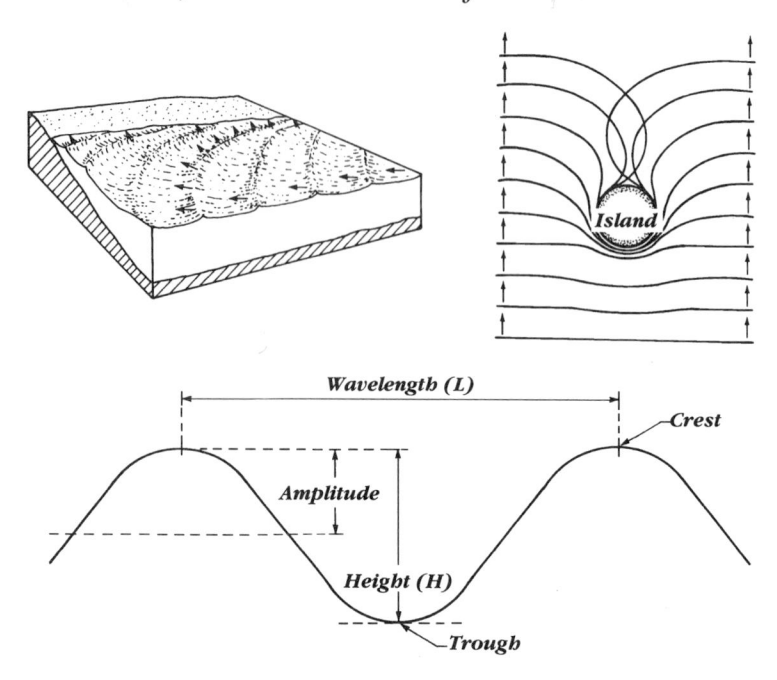

Wavelength (L), *Crest*, *Amplitude*, *Height (H)*, *Trough*

Upper: wave refraction onto shore; middle: wave refraction around island; lower: wave terminology. (Courtesy Henry S. Parker, Southeastern Massachusetts University.)

on a smooth wave, rising to the crest then running down the back side. If the water were not rotating in this manner a boat sailing against a wave train would make little or no progress. —It might even go backwards were it not, in effect, acting like a large chip. And there would be little hope of working off a lee shore. This also explains why a boat running with bare poles before the wind does not lose the mileage that might be suggested by the progression of the wave trains.

The motion of the water in a wave is quite different, however, once the depth equals one-half the length (crest to crest) of the wave and it begins to "feel the bottom." With the resulting drag the top of the wave moves faster than the water below, the wave height increases, the crests become closer together and they begin to topple or break. The effect increases as the water becomes shallower. Where shallow water extends some distance offshore, or where there is an offshore bar or a barrier reef, the unwary sailor may get caught in such breaking waves even while still far from the coast. Regaining sea room once caught in the shallow breakers can be very difficult, particularly under sail alone.

Incidentally, realizing that waves drag along the bottom one can explain why they always approach the shore, even where the wind is blowing offshore. Basically, regardless of the direction in which the waves are moving before they "feel the bottom," the greater drag on the shallower end, in other words toward the shore, causes the entire wave to curve toward the coast.

It is not unusual to see long parallel white lines on the surface, running in the direction of the wind. These may be observed in high seas and persist, sometimes accentuated, in fairly brisk winds. They result from the convergence and subsequent sinking of water rotating in a formation we call Langmuir cells. While all very complicated, what is happening is that the wind creates a circulation at right angles to the wind direction and a series of rotational cells develops across the path of the wind. The lines at the surface develop where cells of opposing rotations meet. Along these convergence streaks there can be quite an accumulation of marine life: phytoplankton, zooplankton, drifting larvae, sargassum and perhaps fishes and sea turtles feasting on the concentrated "goodies." Sea birds may join in. Because of convergence it's a good place to look for the Portuguese man-o-war.

Should it be quite calm you may see other lines on the surface, sometimes the delineation of relatively distinct water masses. Though difficult to decipher, they remind us that the sea is not as homogenous as we may have thought. These may represent upwelling, downwelling, tidal flow, outwelling from an estuary, isolated wind effects, and other interactions, all reflecting a very dynamic environment.

Incidentally, the white foam that superficially looks like shaving cream and may occur along the various lines on the surface or may wash up on the beach, is not as ominous as it might seem. To be sure this may result from the organics of pollution but, on the brighter side, such "lather" commonly represents healthy conditions as life in the water exudes proteins which may mix with air and appear as a froth.

For open voyages the Beaufort scale is commonly used to express the velocity of the wind and the condition of the sea surface. In its original form, as established by Admiral Sir Francis Beaufort in 1806, the scale was primarily linked to the performance of a typical early nineteenth century man-o-war. Forces 1 to 4 told of the resulting speed of a ship. Numbers 5 through 11 related to how much sail could be carried. Beaufort 12 applied to hurricanes when a ship must run under bare poles. There have been modifications through the years, especially in elaborating on descriptions of the sea surface and the effects seen on land. The table included herewith is an up-to-date version.

In a slightly critical but not unappreciative vein, I note that

Winds, Waves and the Beaufort Scale

the Beaufort scale lacks a pattern as the ratios of velocities from number to number are not consistent. And one might wonder why deep sea sailors find it so meaningful, now that anemometers are widely available and weather reports of the wind speed are given regularly in miles per hour? Does this simply reflect a fascination with tradition? Not when we realize that it is useful at sea to describe the effects of the wind as well as reporting wind velocities. Obviously the correlations of wind strength and sea surface effects are most consistent where the wind is almost the sole influence on sea surface conditions. Thus the scale applies best well out to sea and it is used most often in accounts of long distance ocean voyages.

Accordingly, many who are sailing the high seas and referring regularly to working anemometers still use Beaufort numbers, largely to add information about sea conditions. Furthermore, an experienced sailor can generally offer an accurate scale number without checking with an anemometer. Thus, when such an observing sailor logs winds of Beaufort 4, we know that small waves are growing longer with fairly frequent white caps, and that the driving force is from winds in the 13–18 miles per hour range.

Incidentally, as you read the Beaufort Scale, you will see that there is no attempt to correlate wave height with wind velocity. In the few instances when I venture such comparisons I realize my suggestions are open to question. Too many factors, such as wind duration and direction, fetch and complicating currents affect wave height, whereas the sea surface conditions described do relate to the wind. In all of this it is useful to bear in mind that the increasing force exerted by the wind increases approximately as the square of the velocity.

Beaufort Wind Scale

Beaufort number or force	International description*	Wind Knots	Speed MPH	Effects observed far from land	Effects observed on land
0	Calm	Under 1	Under 1	Sea like mirror.	Calm; smoke rises vertically.
1	Light air	1–3	1–3	Ripples with appearance of scales; no foam crests.	Smoke drift indicates wind direction; vanes do not move.
2	Light breeze	4–6	4–7	Small wavelets; crests of glassy appearance, and not breaking.	Wind felt on face, leaves rustle; vanes begin to move.
3	Gentle breeze	7–10	8–12	Large wavelets; crests begin to break; scattered white caps.	Leaves, small twigs in constant motion; light flags extended.
4	Mod. breeze	11–16	13–18	Small waves, becoming longer; numerous white caps.	Dust, leaves and loose paper raised up; small branches move.
5	Fresh breeze	17–21	19–24	Moderate waves, taking longer form; many white caps; some spray.	Small trees in leaf begin to sway.
6	Strong breeze	22–27	25–31	Larger waves forming white caps everywhere; more spray.	Large branches of trees in motion; whistling heard in wires.
7	Near gale	28–33	32–38	Sea heaps up; white foam from breaking waves begins to be blown in streaks.	Whole trees in motion; resistance felt in walking against wind.

Beaufort number or force	International description*	Wind Knots	Speed MPH	Effects observed far from land	Effects observed on land
8	Gale	34–40	39–46	Moderately high waves of greater length; edges of crests begin to break into spindrit; foam is blown in well-marked streaks.	Twigs and small branches broken off trees; progress generally impeded.
9	Strong gale	41–47	47–54	High waves; sea begins to roll; dense streaks of foam; spray may reduce visibility.	Slight structural damage occurs; slate blown from roofs.
10	Storm	48–55	55–63	Very high waves with overhanging crests; sea takes white appearance as foam is blown in very dense streaks; rolling is heavy and visibility reduced.	Seldom experienced on land; trees broken or uprooted; considerable damage occurs.
11	Violent storm	56–63	64–72	Exceptionally high waves; sea covered with white foam patches; visibility still more reduced.	Very rarely experienced on land; usually accompanied by widespread damage.
12	Hurricane	64 & over	73 & over	Air filled with foam; sea completely white with driving spray; visibility greatly reduced.	Very rarely experienced on land; usually accompanied by widespread damage.

*As adopted by the World Meteorological Organization.

Hurricane Winds and Seas

"A tropical storm is developing in the Atlantic, somewhere to the east of the Caribbean." How often have you heard this? And what follows? As the days pass you keep up with the advisories. If the storm develops into a hurricane you follow its track; keep posted; and, if you are in the projected path, you prepare accordingly. But what if you are living on an Eastern Caribbean Island or are sailing in the area? These storms generally develop between 5° and 15° north and may be originating not far away. With a typical trajectory aiming for one of the islands, the prospects for warnings far in advance are not all that great.

Obviously the effect of a hurricane varies with its strength. At the minimum hurricane wind velocity, by definition 73 miles per hour or 64 knots and 12 on the Beaufort scale, the sea is completely white with driving spray and the air is filled with foam. Though winds around 75 miles per hour, exerting a pressure greater than 17 pounds per square foot, are of some danger everywhere and may at least cause prolonged electric, telephone and even water supply outages, they are not likely to cause damage to substantial structures. Along the coast, however, even minimal hurricanes, in fact even tropical storms of some 65 miles per hour, may be very destructive when high waves and storm surges are driven by the winds.

Of course all such damage is greatly accentuated with very high hurricane winds, upward of 100 miles per hour. Such

Hurricane Hugo heading toward Charleston, S.C. (Courtesy of the
National Oceanographic and Atmospheric Administration.)

winds, which against a flat surface may exert almost twice the pressure of minimum hurricane winds, create tremendous structural damage and generate a maelstrom of flying debris. With these effects, plus a storm surge, almost everything along the coast may be leveled.

The storm surges that cause such havoc are partly due to a rise in sea level in response to the low overlying air pressure of the hurricane system. To a greater degree, however, a surge builds as the result of the wind stress on the sea surface and this effect increases at an even more rapid rate than the force on a flat surface as mentioned above. The stress drives the water onto the shallows or into the confines of embayments and estuaries. The greatest damage results from a combination of an unusually high tide, a surge into a confined embayment, and the lasting impact of high wind-driven waves breaking with enormous force in the shallows.

When under power and in control of their steerageway, large freighters, tankers and cruise ships fare reasonably well when caught in the open ocean in hurricane winds and the accompanying seas. Their mobility helps if there is enough information to enable skippers to dodge the worst of the storm. In fact, such large ships are sometimes better off at sea than in port riding on an anchor that may not hold, or tied to a dock, being battered and at the mercy of flying debris. When the harbor setting is questionable, naval ships may be ordered out to sea to ride out a hurricane.

But the story is quite different for smaller boats. The expression "if in doubt head for deep water," applies when in shallow waters near a lee shore, or in a harbor where trouble threatens due to poor holding ground and other unsafe conditions. With the threat of a hurricane, however, the urgent need is for a harbor with nearly complete protection—a so-called "hurricane hole." The safety of these holes varies. English Harbor in Antigua is a near-perfect example, having a narrow entrance, being nearly surrounded by high hills, and providing anchoring space that is generally ample. The expression "generally ample" is used advisedly for, when countless boats head for a favored spot, overcrowding is inevitable. Another precaution, when lacking better protection, is to run small boats far up into the mangroves which may moderate the seas.

So what are the guidelines for the small boat skipper? From early June through mid-October (a bit earlier and later if extra caution is desired) avoid long distance voyages and never stray far from a well protected harbor. Furthermore, don't overlook the danger of a tropical storm with its winds approaching 65 miles per hour and the likelihood of very high waves.

Such precautions are fine if you are willing to stay near suitable protection for more than a third of the year. Needless to say, both of necessity and for pleasure, there is considerable sailing in the Caribbean during the hurricane season, usually with due attention to warnings and avoiding long open-water voyages.

To understand what one experiences in a Caribbean hurricane it is useful to get a bit technical. Various interacting processes "drive" the system. High winds with heavy rains flow toward a low pressure area of warm, moist air rising from the surface of the tropical ocean. The rising air supplies the moisture for the rains around the core while the condensation involved releases heat that accentuates the uplift of air in the central region. The outflow of air in the upper atmosphere at the top of

126 Hurricane Winds and Seas

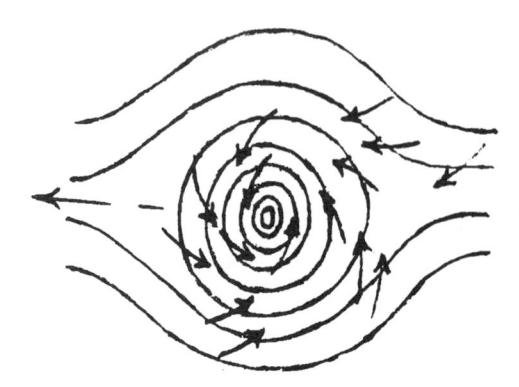

Diagram of hurricane systems. Note the counter-clockwise rotation of the winds toward the center. In this diagram of a hurricane that might be moving west or northwest in the Caribbean, the wind would be much stronger in the northern than in the southern sectors.

the rising column spirals clockwise, compatible with the Coriolis effect* that causes winds to veer to the right in the northern hemisphere. The incoming air spirals in the opposite direction with the result that, at the surface, there is a tight counter clockwise circulation surrounding a calm or "eye."

Starting with a relatively minor tropical disturbance, the system builds as a tropical depression, becomes a tropical storm, and finally a hurricane—which can be mighty fierce as I describe in my write-up of Hurricane Hugo which follows. It is useful to bear in mind that, on the left side of an approaching hurricane, the speed at which the whole system is moving is

*The Coriolis effect results from the rotation of the earth. In the northern hemisphere, flowing air and water are deflected to the right. In the southern hemisphere, the deflection is to the left. You may observe this by watching the water spiral as it drains from a sink or wash basin. I have entertained myself by comparing the drain pattern at home to that seen when visiting in the far South Pacific. Since there are complicating influences in such miniaturized set-ups, the draining water does not always "behave" according to theory.

Hurricane Winds and Seas 127

added to the velocity of the winds; the opposite is true for the right side. Thus a hurricane moving at 20 miles per hour and packing winds of 80 miles per hour may have a real wind effect as high as 100 miles per hour on one side and perhaps only 60 miles per hour on the other side of the system.

Obviously, for anyone at sea the objective is to stay well clear of a hurricane system if at all possible. If not feasible, it is desirable to have the system passing to the left. Bowditch (*American Practical Navigator*) discusses indicators to use to judge the track of a hurricane. His discourse, running for about ten pages, is well worth the attention of a sailor in a hurricane threatening predicament and can also be helpful in knowing what to expect from a fixed position on land or in a harbor. Do bear in mind, however, that observations of natural phenomena can be misleading. As to hurricanes, they may stall and may change in direction, speed and wind velocity. Furthermore, trying to interpret the turbulent conditions in the overcast, the gusts and the wrenching of the waves if one is in a small boat, may be almost impossible. Certainly the basic need is to have the best possible weather advisories, and these days such information, by radio, is quite good.

Two added thoughts come to mind in wrapping up this discussion. First, don't be fooled by the very light winds and clear sky if you experience the core or "eye" of a hurricane. Severe conditions will follow close behind. Also, be prepared to experience high waves deflected at right angles to the winds of a hurricane, making it extremely difficult to hold a course that is both into the wind and head on to the seas.

Waterspouts, being whirling systems in miniature, may seem

like midget relatives of hurricanes. But they are not. It is more accurate to think of them as tornadoes forming at sea, though generally less violent. A water spout may develop when the updraft of a cumulus cloud with interacting air currents creates a vacuum. In the tropics unstable, rising air can produce the same effect. The spray drawn upward becomes the spout you may see. It may be only 5 to 10 meters across and may last only 10 to 30 minutes. Usually the air rotates counterclockwise in the northern hemisphere. Though the resulting waves are not extreme, trying to manage a sailboat in the swirling winds is very difficult. They can be dangerous for small boats, but getting caught in such a system is not very likely.

Hurricane Hugo, one of the worst storms of this century, illustrates the power and devastation of a hurricane. Hugo was spawned off the coast of Africa in the eastern Atlantic. The system intensified as it moved slowly westward, smashing into the French island of Guadeloupe on September 17, 1989 with winds of 140 miles per hour at the surface. As the system continued toward St. Croix, Montserrat was hit hard, while some of the islands to the north of the track, including Nevis, St. Kitts and Antigua were spared the extreme winds, though severely buffeted. Hugo hit St. Croix with winds again at 140 miles per hour and continued without much let-up through St. Thomas and the northeast region of Puerto Rico.

According to reports, 11 people were killed in Guadeloupe, 10 on Montserrat, and 5 in the Virgin Islands-Puerto Rico area. On Guadeloupe, with a population of over 300,000, about 15,000 were left homeless. On Montserrat, structures were leveled over the entire island, leaving nearly all of the island's population of over 10,000 without shelter. In St. Croix some

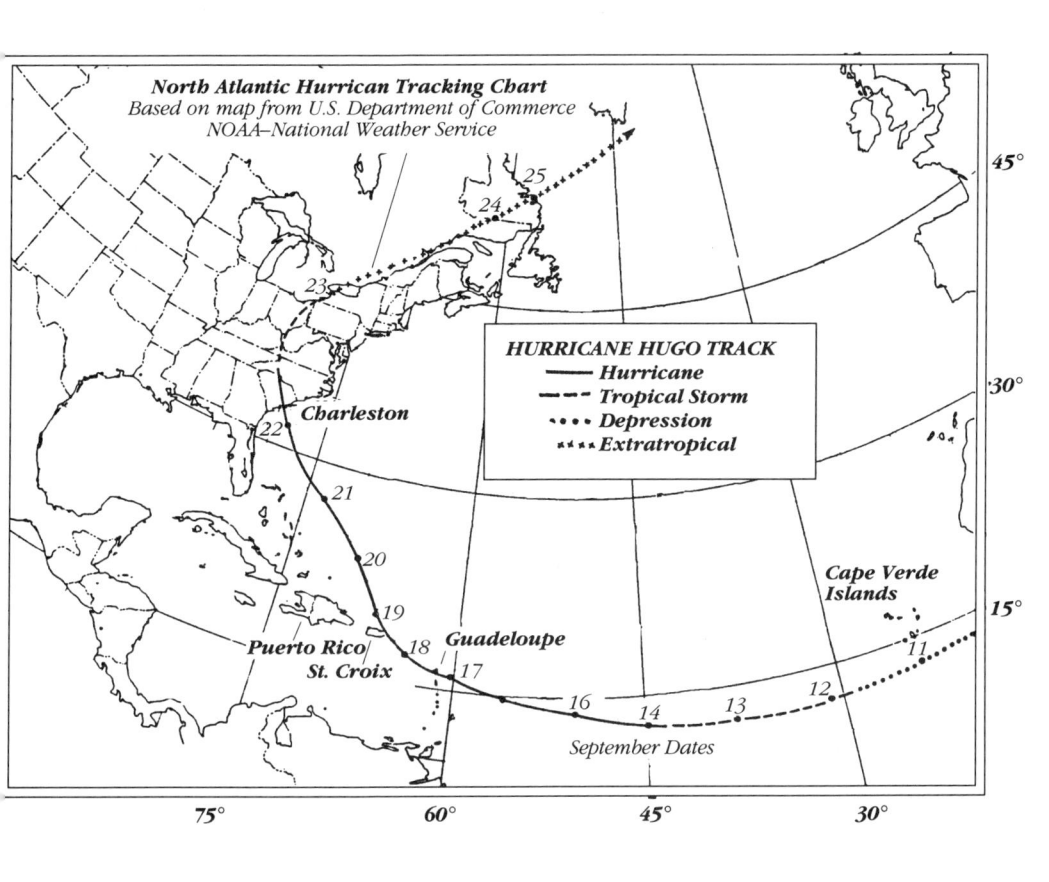

North Atlantic Hurrican Tracking Chart
Based on map from U.S. Department of Commerce NOAA–National Weather Service

HURRICANE HUGO TRACK
— Hurricane
– – Tropical Storm
•••• Depression
✶✶✶✶ Extratropical

Charleston

Cape Verde Islands

Puerto Rico
St. Croix
Guadeloupe

September Dates

50,000 were left homeless. About three hundred boats had been moved to Culebra, an outlying island of Puerto Rico considered one of the Caribbean's best hurricane holes, yet less than a third were still afloat after Hugo had passed. Believe it or not, there is a report of an anemometer reading as high as 170 miles per hour on Culebra.

With St. Croix in a shambles and in a state of unrest from looters requiring over a thousand troops from the United States to control the situation, and with over a billion dollars in damages to Puerto Rico, Hugo left the Caribbean. As is typical of hurricanes, Hugo's winds had lost some of their punch on crossing the land mass of Puerto Rico but, as can also happen in hurricanes, the system regained strength over the open ocean. It hit Charleston, South Carolina, with winds again reaching 140 miles per hour. As a result of the very good tracking and warnings by the U.S. Weather Bureau, many residents retreated from the coast to safer havens inland and, though thirteen lives were lost in South Carolina alone, a more extreme tragedy was averted. But the mix of extremely high winds and waves and a storm surge on the order of 8 feet (well over two meters) above the normal high tide destroyed much of the construction along the coast and left thousands of boats badly damaged and stranded *ingloriously* along the shore and in the marshes.

Hugo's international damage total was in excess of 10 billion dollars according to a U.S. Weather Bureau estimate.

The Bermuda Triangle—Sheer Nonsense

The Bermuda Triangle is known by many names: Devil's Triangle, Hoodoo Sea, Twilight Zone, to name a few. All are suggestive of mystery and supernatural forces. As the persisting legend goes, it is an area of unique dangers, a mysterious threat to those who venture on or over the sea therein. Perhaps, as various explanations suggest, this is a region burdened with more than the usual frequency of water spouts and freak seas. Then there have been suggestions of atmospheric aberrations, magnetic and gravitational anomalies, and strange forces that silence radios, block radar, and affect compasses. More bizarre are the suggestions of death rays from the mythical sunken continent, Atlantis; also the possibility that crews on UFOs have been there collecting earthlings and their vehicles for "study back home." In speculating, some have suggested that there is an enormous wandering abyss within the Bermuda Triangle, engulfing all that it encounters.

The boundaries of the Triangle, as commonly defined, run from Southern Florida to Bermuda to Puerto Rico and back toward Miami. However, since the incidents cited in the legend extend over much of the Greater and Lesser Antilles, I am compelled to deal with it. More to the point—we must put the record straight and dispel the myths. This is especially important as some sixty or so highly publicized disappearances that might seem to defy explanation have occurred in this triangular area. Various writers, some of whom surround themselves with an aura of credibility, keep referring to these and refueling the legend. Such writing is nothing short of shallow, irresponsible sensationalism.

Now let's look at some of the facts, not the myths. First: The number of ships crossing parts of the Triangle is enormous, yet percentage-wise the accident rate is negligible. Commercial and military planes fly over parts of the area every day without incident. The many cruising sailboats making the run from Beaufort, North Carolina to the Virgin Islands every year may experience heavy weather on occasion but nothing unusual or mysterious. (Add us to the list as we wandered through much of the Triangle in 1969–70 with nothing more than the usual challenges that cruising sailors experience.)

Second: Unexplained disappearances of one sort or another are by no means unique to the Bermuda Triangle. After all, when a mishap occurs at sea with no survivors and either without, or at best with confusing radio messages, the event is inevitably full of unknowns. When we say such an incident is shrouded in mystery, we are referring to the unknowns and are not implying any unfathomable dangers. If we continue with the kind of thinking that has perpetuated the Bermuda Triangle legend, we will need another large triangle to surround the disappearance of Amelia Earhart. In fact, if we are to cover in this way all the unexplained tragedies related to ocean travel, we will need triangles all over the face of the globe.

Third: In their zeal to build on the Triangle legend, various writers have included incidents that really occurred elsewhere. For example, the mystery of the *Mary Celeste*, discovered drifting crewless between the Azores and Portugal in December 1872, is frequently mentioned in relation to the Bermuda Triangle. The sails were set and things on board the deserted craft were basically in good order. She was sailing erratically with the wind, having progressed 500 miles to the east in the eleven days since the last entry in the log, 100 miles west of the Azores. So the *Mary Celeste* was at least 2,000 miles east of the Bermuda Triangle when abandoned but, being such a mystery, reports linked it to everything imaginable including the Triangle.

Finally: As serious oceanographic and related research has taught us more about this region of the ocean and the air above, we have not uncovered anything unusual. If we had doubts, would we launch our satellites right into the heart of this Triangle?

One writer, Lawrence Kusche, became sufficiently impressed with the stories on which the Bermuda Triangle legend has been built that he reviewed some fifty incidents identified as indicative of unique dangers. For the great majority, very plausible explanations unfolded. His most extensive review dealt with the flight of five Navy avenger torpedo bombers lost on a routine patrol over the Bahamas in 1945. Probably no other casualty did more to add to the legend of mysterious dangers within the Triangle. And probably no event that occurred in the area has been studied more exhaustively. In its report of more than 400 pages, the Navy Board of Investigation listed fifty-six "facts" and fifty-six "opinions." The Board made reference to the fact that the flight leader had failed to change his radio to the

134 *The Bermuda Triangle—Sheer Nonsense*

emergency channel to clarify questions of the flight's true position.

As I write this, there are news and TV reports that the five planes from the training flight of 1945 had been found on the sea floor east of Fort Lauderdale. Soon there were denials; the planes found were from a different flight. No doubt the legend builders will play up this discovery and denial, paying little attention to Navy records of numerous losses elsewhere of this same type of aircraft.

In the concluding remarks of his exhaustive study Kusche says:

> The legend of the Bermuda Triangle is a manufactured mystery. It began because of careless research and was elaborated upon and perpetuated by writers who either purposely or unknowingly made use of misconceptions, faulty reasoning, and sensationalism.

I conclude on a light note to the effect that undue alarm can lead to humorous situations. Supposedly true is the story of a neophyte skipper on a bare-boat charter who, from an anchorage on the edge of the Triangle, radioed his charter service greatly alarmed that he had been caught in a mysterious vortex. "Is your anchor still down?" came the reply. "Oh, thanks; we'll be all right now!"

The Tropical Sky

Trying to catch a glimpse of the unique "green flash" from the sun as it is setting is a real challenge when you are traveling in the tropics. At these latitudes, the flash lasts only a fraction of a second. It can only be seen if the atmosphere is crystal clear in your line of vision to the horizon where the sun is setting. It occurs just as the sun drops below the horizon when the atmosphere, acting as a prism, separates the colors of the sunlight. The long wave lengths of the red and yellow end of the spectrum are refracted the least and are the first to disappear below the horizon. Blues and violets are scattered by atmospheric particles, leaving the green visible. Less likely, yet possible, is a green flash at dawn. If you happen upon sufficiently clear conditions in more northern latitudes, the green may appear for minutes instead of split seconds.

Even in the rainy season, from June until early November, one can expect a good deal of clear weather and bright skies over the Caribbean. Cumulus clouds build up overhead, and often in the distance one sees dark rain squalls descending from these clouds. The approach of a squall bears watching. It may bring a wind shift and a sudden blow. On the more positive side, the freshwater will surely be welcome by the small boat operator with a limited supply. High clouds above the cumulus

136 *The Tropical Sky*

may suggest weather developments, such as fronts, etcetera, to the north.

Occasionally there is a haze high overhead as dust is blown from the Sahara across the low latitudes of the Atlantic. If at all pronounced, which is more probable from May through October, the dust is likely to appear reddish brown. The lesser haze of other seasons may be more grey or black. I am told that locusts from plagues in Africa have been recovered in the Caribbean. Paul Joyce of the Sea Education Association says they have picked up live ones in their nets used to sample the sea surface. The common advice around the islands is to *kill* any locusts that one encounters.

In the Caribbean one may enjoy some 11 hours of sunshine during the months when daylight decreases to not much more than 9 hours in the north temperate areas. This is due, of course, to the tilting of the earth until, on December 21 and 22, the sun's path is over the Tropic of Capricorn (23 ½°S). Perhaps more than we realize, the longer days are a major plus factor for those who want to retreat in winter from the short, gloomy days of the temperate regions to the longer days and the fair weather of the Caribbean dry season. The situation is reversed somewhat when, by June 21, the sun is over the Tropic of Cancer (23 ½°N) and sunshine may last for 15 hours or so in mid-latitudes; yet even then there are about 13 hours of sunlight in the Caribbean. Incidentally, with the sun generally rising and setting at less of an angle in the tropics than in the temperate latitudes, twilight tends to be brief. Though darkness sets in quickly once the sun goes down, one can count on some memorable sunsets.

Since the Eastern Caribbean is in a belt some 15° east of the

The Tropical Sky   137

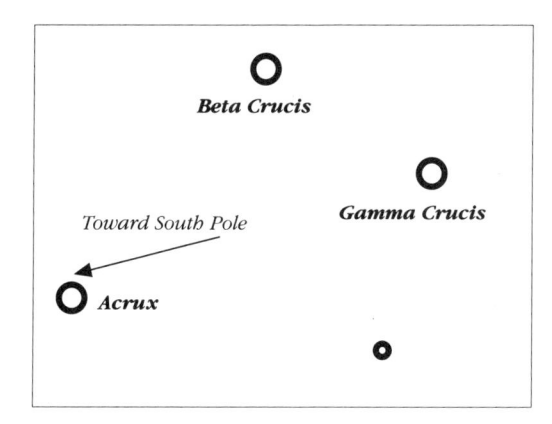

The Southern Cross as seen from the Caribbean in the early evening in February.

eastern United States, the sun rises and sets earlier. Accordingly, the islands are under Atlantic Standard Time with clocks set one hour earlier than Eastern Standard Time. In the seasons of longer daylight there is a shift to a Daylight Savings mode in the Eastern time belt, advancing all clocks by one hour, so the one-hour difference does not apply.

When the weather and skies are clear, the viewing of the stars and planets is especially good, and there are some superb, beautiful moonlight nights. If you are familiar with the stars as seen in the north temperate latitudes, the scene overhead will, with a few adjustments, still be familiar. The North Star, for example, is very near the northern horizon and all constellations are shifted a bit to the north. At these latitudes it is useful to have a star guide for the Southern Hemisphere since, looking far to the south, you can begin to pick up some of the characteristic stars that are overhead "down under." A common goal is to spot the Southern Cross, often seen from as far north as the northern Caribbean. Centaurus, a constellation near the Cross, has two first magnitude stars, Kentaurus and Agena. Also to be seen toward the South Pole are two separate galaxies, the Small and the Large Magellanic Clouds, which are close companions to our own galaxy, the Milky Way.

A major bonus in watching the night sky is the chance to get away from *light pollution*. Well out at sea or in a quiet, secluded anchorage, the night viewing may be the best you have ever experienced. Under such conditions I enjoy watching the shooting stars and waiting for a satellite to cross the sky. The latter do appear frequently. They are easily distinguished from the short-lived shooting stars, but are not as readily differentiated from airplanes flying at 25,000 to 40,000 feet.

The Indies of Columbus

Did Columbus discover America? For the people of Western Europe—Yes. For Mankind—No. These continents were populated long before Columbus first set foot in the Caribbean. Some of the early native civilizations were very advanced and sophisticated. Also, as is widely known, the Norsemen had made numerous landings and had established temporary settlements to the north. However, the people of the western European countries, who were destined to prevail in the Western Hemisphere with their commerce, civilization and culture, were virtually unaware of all this.

Long ago, western scholars had realized there was something to be discovered "out there to the west." Aristotle is said to have written that the crossing from Spain to the East Indies could be made in a few days. As early as the time of Christ, voyages far to the west had been undertaken but without significant discovery. Ultimately, curiosity concerning the sea to the west grew with the persisting expectation that, by sailing far enough, explorers might find a more favorable trade route to Asia, particularly to the East Indies. Estimates of the distance westward to Asia were varied and typically less than a third of the actual mileage. Columbus was especially optimistic with his figure of 2,400 miles* from the Canary Islands to Japan (then called

*The actual airline distance is 10,600 miles.

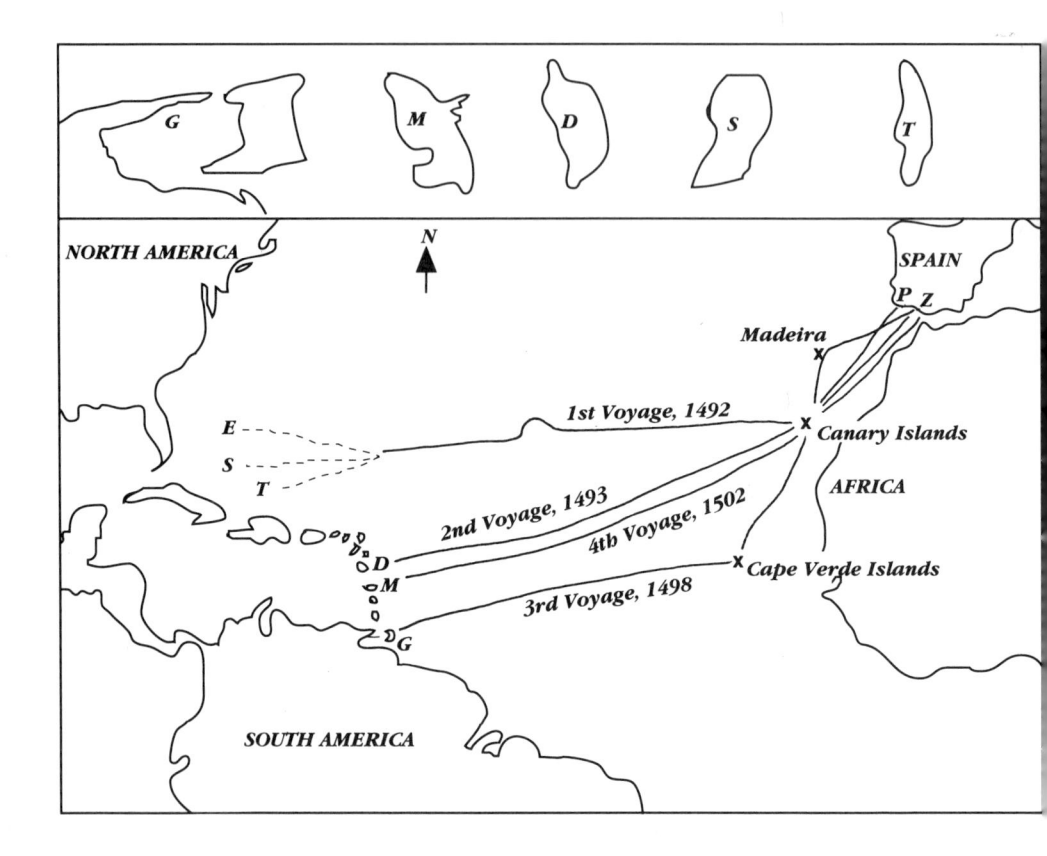

Voyages of Columbus to the West Indies: P = Palos; Z = Cadiz; E = Egg Island in the Northern Bahamas; S = San Salvador; T = Grand Turk; D = Dominica; M = Martinique; G = Gulf of Paria with Trinidad at the mouth.

Cipangu). This helped him to gain support for his voyages from Ferdinand and Isabella, king and queen of Spain. After all, Columbus did get to the Indies, to the West Indies that is.

The fabled island referred to as Antilia was supposedly on the route from the Canaries to Japan. According to Toscanelli, a leading geographer of the time, it was situated about a third of the way to Japan, and Columbus thought it might be a good stopping point. Though there is no such island, it seems that the name persists as we refer to the various eastern islands of the Caribbean as the Greater and Lesser Antilles.

Various historians, navigators and sailors, plus a few oceanographers have tried to identify the island where Columbus first landed in 1492, a site he named San Salvador. For some years, capable scholars pointed to Watling Island in the Bahamas as this historic landfall and two, who have attempted to retrace the voyage in their own boats, namely Robin Knox-Johnston, famous as the winner of the first solo round-the-world race, and the historian Samuel E. Morison, add credence to the Watling Island landfall. Apparently the British endorsed this some time ago when they added the name San Salvador to Watling. Widely referred to currently is an account in the November 1986 issue of the *National Geographic* pointing to Samana Cay, about 75 miles southeast of Watling/San Salvador, as the landfall.

Roger Goldsmith and Phil Richardson of the Woods Hole Oceanographic Institution have probed the landfall question exhaustively and have summarized (though they have not as yet published) their interpretations. Talking to Phil, a former University of Rhode Island faculty colleague, I sense that it is impossible to specify the landfall because of uncertainties in the length of the league as applied by Columbus and the field of magnetic variation in 1492. As a new twist in seeking the answer, Goldsmith and Richardson have been probing these unknowns by analyzing other voyages of Columbus for which the destinations are known, for example reaching the island of Santa Maria on the return voyage and the landfall at Dominica on the second voyage. Depending on the figures applied, the landfall possibilities range all the way from Grand Turk to Egg Island, the latter being well to the north in the Bahamas. Their emphasis on Grand Turk seems a bit unique as these two oceanographers apparently consider it a strong possibility. Of course, except for celebrating, it doesn't make too much difference where Columbus landed. If I were to celebrate, I would just pick my favorite spot and challenge anyone to prove me wrong.

From his first landfall Columbus worked his way south. He explored the northeast coast of Cuba, then returned to Hispaniola. After various contacts were made along this coast, his flagship, *Santa Maria*, was wrecked on a coral reef. While preparing to return to Spain on board the *Niña*, Columbus set up a fortified camp of some forty men on the north coast of Hispaniola. He named this *Villa de la Navidad*.

Following the outgoing voyage to the Bahamas under the relatively steady east and northeast tradewinds, the prospects of a return voyage into the wind must have seemed pretty grim. Actually Columbus took a route quite a distance to the north into the westerlies which were favorable, though too strong at times. Historical accounts differ as to whether this return route was preplanned or was more of a happenstance. It is also not clear whether, having experienced this advantageous eastward passage, Columbus realized that it might help to return this way regularly in the future. It would seem that he did not appreciate the route he had happened upon as he and his contemporaries stayed more to the south on some of the succeeding trips. This must have been very tedious in the ships of those days that could sail no closer to the wind than 67°, as compared to 45° for modern craft.

For Spain, it was a different world after the first voyage of Columbus. Expectations as to the resources of the "discovered" land ran high and the sovereigns were bent on taking advantage of this. The second voyage of Columbus involved a large flotilla with eager participants and ample supplies to strengthen their first settlement in the Caribbean. From then on support and follow-up sailings by others became increasingly common in gaining a European foothold in the region.

On the second voyage, Columbus first landed on Dominica. He continued onward only to find that La Navidad had been wiped out. The colonizers had feuded amongst themselves and had abused the natives who attacked and butchered them. Columbus continued his two-fold mission, both to explore and to found a Spanish trading post. A new settlement called Isabela was established, not far from the Navidad site. As a colonial administrator, however, Columbus experienced considerable difficulty. In fact, at the end of his third voyage, which involved touching the American continent for the first time on the coast of the Gulf of Paria west of Trinidad, Columbus was sent home in chains by an envoy who had investigated abuses and mismanagement in the Hispaniola settlement.

Columbus recovered his status at least to the extent that the king and queen, who were beginning to direct their support for colonization to others, enabled him to make one more voyage. It has been said that the king was sort of glad to get Columbus out of the way while the queen still had a lingering faith in his explorations, or at least had an uneasy conscience over the way he had been treated. His fourth landfall was Martinique, and he went on as far as the coast of Central America and the northwest coast of South America. All told, the explorations of his four voyages were fabulous, yet Columbus died thinking he had reached the eastern shores of Asia.

The voyages of Columbus were in marked contrast to the earlier exploits of the Norsemen who, step by step along the shorter, great circle route (via Iceland, Greenland and Newfoundland) had found their way to southern New England. They had made repeated voyages and had set up colonies in North America. Apparently they also wandered far inland. On the other hand, their motivation for trade and riches was not comparable to that of the European countries to the south and they did not sustain their foothold. One may ask why Norway or Iceland did not press its rights to lands in North America when, after the discoveries by Columbus, the Pope was proclaiming that the Spanish Crown was entitled to "all heathen lands and islands" already discovered west of the Azores. Simply put, it was it was well into the sixteenth century before geography had linked northern and southern discoveries as relating to one and the same large continent.

We usually think of the period following the voyages of Columbus as an era when the Europeans were discovering America, typically sailing west on the tradewinds and returning in the belt of the westerlies once the advantage of this circle route became apparent. As Spain was establishing itself in the Caribbean, neighboring European countries were not sitting idly by. To some extent, Spanish claims around the Caribbean were contested but the major ventures of England and France were more to the north. Portugal's major thrust was in the eastern part of South America.

From another perspective, this was a period when the Europeans began to "move in on" the native Americans. The interactions of the cultures inevitably led to domination on the part of the Europeans. In the Eastern Caribbean the Indian populations simply succumbed to the power (guns and all) and transmitted diseases of the white man. Basically, the explorers and early traders wanted to exploit the native riches, mostly gold, which was not in great supply in the eastern islands, though an abundance was later found in Mexico. The so-called "savages" of the Eastern Caribbean were regarded as inferior humans needing to be converted to Christianity. The situation became even more tragic in Mexico, Central America and west in South America where the colonizers, under the glamorous identity of conquistadors, brought an end to the great Aztec, Maya and Inca civilizations.

The continuing interest in the Eastern Caribbean by the settlers from Europe is further developed in the next section titled "History as Influenced by Natural Features."

History as Influenced by Natural Features

The Lesser Antilles served as stepping stones for Indians from the northern coast of South America to populate the Eastern Caribbean. The Tainos, a relatively peace-loving group, were followed by the more aggressive and savage Island-Caribs.* The Tainos were the first to greet Columbus in the Bahamas, Cuba and Hispaniola. His later ventures into the Lesser Antilles involved contacts with the Island-Caribs.

In what we might think of as the immediate post-Columbus era in Caribbean history, the primary use of the islands was for defense and attack in relation to shipping. The Spaniards, who had come upon rich sources of gold, silver and other precious minerals in Mexico and in western South America, were shipping it in galleons sailing through the Mona Passage (between Puerto Rico and Hispaniola), the Windward Passage (between Hispaniola and Cuba), and the Straits of Florida (along the Keys and northward). The routes were dictated by the geography of

*You may often hear the Tainos referred to as Arawaks, though some scholars point out that they differ from the Arawaks of South America. Desmond Nicholson of the Antiguan Archeological Society thinks there may have been Indian populations in Antigua almost 2,000 years B.C. and suggests that the earliest presence of ancient humans may even date back as far as 10,000 years B.C. The Tainos came to the region about the time of Christ, something on the order of 1,500 years before the Europeans arrived.

R.V. *Trident*, former University of Rhode Island research ship used by Tony Sturges for deep current measurements into the Caribbean and by the author for cruises to the Caribbean and Bahamas; and S.S.V. *Westward* of the Sea Education Association, one of two S.E.A. ships on which the author traveled in the Caribbean. (Photos courtesy of the University of Rhode Island and the Sea Education Association.)

the winds as it was important to clear the Caribbean and the tradewinds, then to work north to the westerlies of the temperate latitudes for the return to Europe. Accordingly, island sites in the Eastern Caribbean were fortified by Spain to protect its shipping through the "bottlenecks" of the passes. Other sites were used by outlaw pirates and other raiders from England, France and Holland, some having arrangements with their respective sovereigns to plunder the Spanish fortifications, to raise havoc with the shipping and to profit in the process.

By the seventeenth century, the English had established a major attack headquarters at Port Royal, Jamaica, under the command of Sir Harry Morgan, one of the most famous of the privateers. Late in the century, Port Royal, with its burden of sins, was drowned by an earthquake.

By the end of the seventeenth century, after the competing European powers had sapped the strength of the Spanish and had broken Spain's dominance in the Caribbean, an extended period of agricultural prosperity was being launched, largely by colonial regimes of the English, French and Dutch. This was based on three natural assets:

The soil and the climate were favorable for the growth of sugar cane, cotton, tobacco, coffee and cocoa.

The northeast tradewinds provided free energy to drive windmills for grinding sugar cane which had quickly become the major plantation crop. Conveniently, the trades were strongest during the dry season at the time of cane harvests.

The planetary wind systems also provided free energy for sailing ships that became engaged in a profitable triangular trade. Loaded with sugar, rum and other products of the islands, they sailed the westerlies to Europe. Returning with supplies for the West Indies, they detoured via the west coast of Africa where Blacks were impressed to be sold as slaves for the plantations. The triangle was closed as these ships, with their supplies and slaves, ran down the tradewinds to the Caribbean islands.

Though there were various naval battles in the Caribbean, linked to the warfare between the powers in Europe and though the control of individual islands changed hands from time to time, the sugar plantations and other agricultural ventures prospered. In 1815, when the crowned heads of Europe assembled in Vienna to bring order out of the chaos that had been created by Napoleon's imperialism, the status of the Caribbean colonies was eventually settled. The major changes that followed involved the acquisition of Puerto Rico by the United States in the settlements following the Spanish-American War, the purchase by the United States in 1917 of Denmark's holdings in the Virgin Islands, and the moves by Great Britain to grant independence to its colonies in the last half of the twentieth century.[*]

Subject to the rulings of the European powers, slavery was abolished most everywhere in the West Indies before the middle of the nineteenth century, well before it was abolished in the United States. Under a modified economy, the agriculture continued but the great prosperity of what had become known as the "sugar islands" was a thing of the past. More than a hun-

[*]An update comparing the past affiliations following the Treaty of Vienna and the present status of these islands is given at the end of the opening section on "The Island Arc."

History as Influenced by Natural Features

dred years went by before, well into the twentieth century, a new economic thrust developed based on tourism. The sand, the beaches, the splendid mountains and the climate had become feature attractions. And it was the unfolding convenience of air travel that made the tourist boom possible.

The islands also proved to be a natural setting for cruise ships which can easily make port on one island for a day, then travel overnight to another for guests to go ashore for shopping after breakfast. Similarly, the island setting has proved ideal for yachting, including chartering, since it is so easy to run from one anchorage to another along the western or lee shores. The multiple country jurisdictions complicate this since, from one short hop to another, those traveling on small boats may find it difficult to deal with various entrance and departure routines and accompanying fees.

Many of the people one sees on these islands today, often thought of as natives, are the descendants of slaves. Very few of the true natives, i.e. Tainos or Island-Caribs, have survived. In some places there are quite a few immigrants from India. While language, customs and governmental structures encountered in each island reflect its colonial history, or its continuing dependence as the case may be, the Blacks and Mulattos have some dialects of their own. Much that is unique to those with African blood is shared throughout the region as communications have always been rather effective over the short interisland distances. The steel drum bands enjoyed throughout the islands are an example of shared cultural development. As this is an adaptation to equipment not available early on, it is an impressive expression of a widely shared sense of rhythm.

The more prosperous element of the population includes descendants of former plantation owners. Their numbers are rapidly being augmented by newcomers from Europe, the United States and Canada for whom the natural charm of the Eastern Caribbean is very appealing, especially during cold months in the north.

The Major Resources

Sugar, a luxurious climate, plus fish and a "pinch of salt." —As I discussed in my comments on history, sugar was the basis for the early growth in the economy of the islands while climate, along with other natural features, is the foundation for an economy increasingly based on tourism.

Though the great sugar plantations were linked to the bygone days of slavery, sugar cane is still a major agricultural crop on some islands. Other crops, such as bananas and pineapples, round out the present production.

Judging from the hotel construction underway and contemplated, it appears that planners are banking on the continued growth of tourism. Local labor, an advantageous resource in terms of availability and the modest wages expected, is used increasingly in support of tourism. For labor this helps take up the slack as developments in agriculture are comparatively limited and mechanization tends to substitute for manpower. Under such circumstances, it is all too easy to fail to realize that the natural assets, not tourism per se, constitute the resource involved and that any abuse of natural features can undermine a tourist-based economy. This is discussed more fully in the write-up of "Conservation."

The prospects for living resources from the sea were discussed in the write-up of "Fisheries," with a few comments on the added prospects under "Aquaculture." While fisheries are very important at the artisan level, there is little likelihood of major development other than for sports fishing linked to tourism.

We must not forget the "pinch of salt." Actually salt production from evaporation ponds, though important in some locales, no longer ranks as a major resource.

With the construction needs that accompany development, sand and gravel rank as major resources. Of course there's plenty of it. Unfortunately, the sites where mining can be done most economically are along the shore and, *very unfortunately*, this is sometimes done in defiance of laws and restrictions. With due provisions to safeguard the landscape some sand and gravel can be taken from upland sites. Mining sand offshore is also feasible, though it involves added operating costs. To do this without secondary damage to the coastal features requires considerable knowledge of the local underwater hydrography.

Most petroleum resources have been found in the southern Caribbean, so the expectations for these non-living resources are not great, except in Barbados and Trinidad and Tobago.

Law of the Sea and the Island Countries

International agreements of special significance in protecting the marine environment of the Caribbean include the Cartagena Convention plus certain widely accepted provisions of the Law of the Sea Convention.

The Cartagena Convention, drawn up in 1983, went into force in 1986 as the mechanism for achieving the Caribbean goals of the Regional Seas Program of the United Nations Environment Programme. The agreements address the responsibility of Caribbean nations to control marine pollution from land-based and atmospheric sources and from dumping and seabed activities. Though most nations in the Caribbean region have approved the Convention and the goals are the subject of follow-up conferences and monitoring efforts, setting up specific plans that are both attainable and sufficiently restrictive is proving difficult. Oil pollution is considered to be the priority environmental problem of the region.

Whether or not the international Law of the Sea Convention (LOS) promulgated in 1982 is adopted, or whether it becomes accepted in some amended form, many countries of the Eastern Caribbean have accepted and comply with most of its provisions. Thus there is almost universal adherence to the provision that countries can claim rights to the resources within an area

Law of the Sea and the Island Countries 155

extending 200 miles from the shoreline. With such claims the states accept binding obligations to protect and conserve these resources. The extended area is referred to as an "Exclusive Economic Zone" (EEZ).

For the present, most of the resources of the EEZ of the Eastern Caribbean are fisheries. Any enterprise from abroad that may wish to harvest fish within the 200-mile area must receive permission and be suitably licensed. For Puerto Rico and the U.S. Virgin Islands this is handled through the United States government. For much of the rest of the Eastern Caribbean, licensing provisions have been standardized by the Organization of Eastern Caribbean States (OECS) which is also developing common surveillance and enforcement procedures. The island countries which are not independent but are colonies or departments of European countries cooperate with the OECS.

One LOS provision that is widely applied is that of requiring permission for any research to be carried out within the 200-mile extended limits. Commonly this involves accommodating each country by taking one of its observers on board the vessel from which the research is conducted, hopefully someone who can be an effective contributor to the research. Permission commonly involves reasonably prompt and shared release of all data obtained. While this concept of cooperation merits respect, such clearances are obviously something of a deterrent to research in view of the many small countries that must be approached before conducting studies off the islands of the Caribbean.

Of considerable interest in relation to transportation and military operations is the provision in the LOS for the "right of innocent passage" through the territorial seas of coastal states, which generally extend 12 miles offshore.

Countries may establish regulations and liability provisions for pollution violations and accidents within the boundaries of their territorial seas. To a lesser extent, and particularly when the conservation of resources is involved, states may deal with such issues over their entire EEZ. Generally, however, matters beyond the territorial waters are high seas issues dealt with by the International Maritime Organization (IMO). Fortunately IMO is improving its rulings on disposing of wastes, cleaning ships' oil tanks, etcetera. One of its latest provisions is a total ban on dumping plastic waste at sea. While most nations endorse the IMO protocols, you can well imagine that enforcing adherence to some of these rulings is difficult.

Referring back to the 200-mile zones, you can sense immediately that there are broad overlaps in the offshore claims of the island countries of the Caribbean. Progress is being made in negotiating acceptable boundaries and the concern is not nearly as great as elsewhere in the world where the prospects for oil are substantial and highly contested fisheries are at stake.

Conservation

Though the need is appreciable, conservation measures seem rather minimal through much of the Eastern Caribbean. We enjoy the beautiful reefs; yet in some areas there is inadequate control of the picking and gleaning for coral souvenirs.* Fishing is a way of life for many along the coast; yet in some areas there is a need for increased attention to fisheries management and enforcement. Then there's the assault on the beautiful beaches, marred here and there where sewage runs in open rivulets to the water's edge. And, as to trash, the mess sometimes seen along the roadways makes one wonder how people so dependent on tourism can be so careless.†

One may wonder why I emphasize the negative while all the while I so enjoy the Caribbean setting. We benefit from the general resilience of the tropics that often tends to override the degradation. But it is this resilience that can blind us to what is happening and cause us to overlook situations where habitats may be in a state of near collapse.

We must also be aware that oceanic islands, chiefly because of their small size, are especially vulnerable to man's intrusions. Generally the lowlands and the coast are under the greatest

---

*Unfortunately various regulations as to what can be exported from the islands, in commerce or by tourists, are not always enforced.

†I know of areas in the United States that are every bit as bad.

Upper: spotfin Butterflyfish (courtesy C. Lavett Smith, American Museum of Natural History.) Lower: reef Crest with Elkhorn Coral and Sea Whips (courtesy of James R. Sears, Southeastern Massachusetts University.)

stress from our use; the wet highlands are less affected. As areas are cleared for agriculture and development, some environments essential for the unique life of the tropics are disturbed.* While I am concerned regarding negative aspects of man's impact, I do recognize that some of the net results are beneficial. After all, sugar cane, bananas and other introduced agricultural crops have sustained the people and supported the economy. Also, much of the introduced flora that has prospered is very attractive, even though it often dominates over the endemic vegetation.

Probably nothing is less appreciated in the natural setting than the extensive mangrove tracts and probably nothing is more appreciated, yet less cared for, than the coral reefs. Both are very vital to the stability and protection of the shoreline: the mangroves in holding sediments and building land where it would otherwise wash away; the reefs acting as a buffer against the eroding force of the waves. The seagrass beds in between also trap and stabilize sediments. Furthermore, these environments are vital to the productivity of the region. The mangroves and seagrasses are key nursery and nourishing environments and the corals provide the "home" for many important fishes.

Few natural complexes suffer so much from man's abuses. There is a dire need for erosion controls to counter freshets and silt run-off, for advances in sewage treatment, for reduction of oil damage from passing shipping, an end to filling mangroves for development, and a substantial reduction in damage to reefs whether from tourists or fishermen. You can see that the challenge to conservation is staggering and, if we are candid, we

*Of course, loss of natural habitat is by no means unique to the tropics.

must admit all is not going well. Yet, where well cared for, there is nothing healthier in nature than a coastline where luxuriant corals, seagrass lagoons and extensive mangrove tracts interact.

Some environmental setbacks remain a mystery. In 1983 there was an unexplained die-off of the long-spined sea urchin throughout the Caribbean. As mats of algae began to grow, where they were normally held in check by these herbivorous urchins, there was concern that they might smother the corals. However, such worries vanished when the long-spined urchin population recovered as mysteriously as it had vanished.

The next mystery occurred in 1987 and again in 1990 as the one-celled algae, the zooxanthellae, were expelled from the tissues of the corals. This "bleaching" and the possibility of serious consequences are discussed in the write-up of "Coral Reefs."

Tourism presents a challenge to and leads to some conflicts with conservation goals. While the natural features of the area are essential in the appeal to tourists, it is difficult to accommodate the desired influx without impinging on the environment. Obviously the gold coasts of our planet, of which the shores of San Juan, Puerto Rico are a prime example, often wipe out vast mangrove areas and threaten the beaches and reefs. Recognizing that they serve a need in terms of facilities and services, about the best we can do is to assure good management in such critical matters as sewage treatment, disposal of wastes, runoff control, and so forth.

Probably there is much more to gain by overseeing future development, not only with environmental regulations but also by encouraging wise and imaginative steps wherein development and conservation may be reasonably compatible. Finally,

Conservation 161

since development is inevitable, it is important to seek a balance with conservation, in other words, the greater the development effort, the greater the need for matching measures including parks, reserves, use restrictions, and mitigation measures such as reforesting abandoned fields. The list of helpful measures that might be pursued could go on and on. Some may be costly but not nearly as much so as a collapsing, unattractive environment.

Fortunately many leaders are aware of the conservation needs. In the area-wide perspective one can look to the United Nations Environment Programme with its Caribbean Environment Program, formulated and supported by most of the countries bordering on that sea. This Program, and the Cartagena Convention designed to carry it out, seeks to address the management of marine and coastal resources, the assessment and control of marine pollution, and it encompasses training, education and public awareness (see write-up of the "Law of the Sea"). Hopefully, endorsements of the Convention on International Trade in Endangered Species of Wild Fauna and Flora (CITES) will become more universal. The prohibitions and controls of this treaty are needed to stop the marketing of tortoise shell, black coral jewelry, and other products made from endangered or threatened species.

Island to island there are a number of specific conservation laws and practices. In general there are efforts to regulate against abuses in mining the sand and gravel used in support of the construction industries. The protection of migratory birds receives some attention but more is needed both from the standpoint of hunting regulations and habitat conservation.

162 Conservation

While there may be considerable hunting for mammals in the highlands, particularly in the heavily wooded regions, reasonable limits usually apply to protect such species as agoutis, opossums and raccoons. There are various parks and reserves, though some exist more on paper than in reality. In fact a protected reef with an abundance of large, relatively tame fishes seems hard to find. Hopefully, with an increasing awareness, we can look forward to better management and enforcement.

Most every island country has restrictions against taking turtles during their breeding seasons, particularly when they come up on the beach to lay their eggs. Seeking to protect the eggs from predation is another common goal.

Protecting spiny lobster populations is a widespread objective, usually through harvesting bans of several months. For optimum effectiveness in fisheries conservation, increasing attention is being directed to working out common management provisions from island to island as is logical for dealing with creatures oblivious to separate island jurisdictions. The intergovernmental coordination now being worked out should also enhance the control over outside nations partaking, sometimes in excess, of resources in the economic zones that extend 200 miles offshore of these countries as discussed under "Law of the Sea."

A Pollution Wrap-Up

Over and over in these write-ups, pertinent comments on pollution come to the fore. A few generalities seem in order at this point.

The most serious pollution in the Eastern Caribbean is from external sources, such as bilge pump-out from passing tankers, occasional tanker wrecks, plus spills at the few island sites where oil refineries are located. Oil can smear the beaches and suffocate life in intertidal and shallow areas including coral reef communities and mangrove roots. As a consequence of tanker traffic patterns, petroleum wastes are likely to be worse on the windward than on the leeward beaches. Similarly, plastics and other trash can also be especially annoying on windward shores.

The most widespread problem originating in the islands is freshwater runoff which, with the accompanying silt load, can be very damaging to marine life. The problem is greatest where vast areas have been deforested, sometimes for lumber and wood for cooking and, on a major scale, to clear land for agriculture. Without accompanying erosion control, the resulting stress is accentuated when heavy rains fall on the barren terrain. The freshets that may result can extend many miles off and along the shoreline.

Perhaps from the standpoint of health hazards and obnox-

164   A Pollution Wrap-Up

ious odors, domestic pollution and trash should not be third on this list. Especially troublesome are the sites where outfalls and run-off result in excessive nutrient loading, ecologically known as accelerated eutrophication. This can lead to undesirable growths such as mats of algae smothering coral reefs. Also, the rotting of increased plant growth can lead to localized oxygen depletion. With the high ratio of surrounding sea to populated land the old adage, "the solution to pollution is dilution" tends to relieve all but the most local aspects of this problem.

How does this relate to small boats in the area? With some exceptions, most notably in Puerto Rico and the Virgin Islands, the direct discharge from the heads (toilets) is the accepted practice and, where there are restrictions, enforcement is seldom workable since service provisions, such as pump-out stations, are a rarity. The disposal of trash is another story. Containing it for later disposal in port is usually worth the effort, though I watched as a local boat transported garbage to be dispersed in the currents after I had painstakingly brought it ashore.

Large ships are a different matter. Obviously a large vessel, particularly a cruise ship, would create a disastrous mess if its heads were to flow freely when in port. So they have large holding tanks and pump overboard when at sea. Does this sound rather obnoxious? Sure, but dilution can handle it and passenger sensitivity is commonly circumvented by pumping out underway at night when most passengers are asleep. The trash on board may be compacted and incinerated, with some stored until reaching a port with facilities to handle it. Today there is a complete ban on the disposal of plastics at sea from

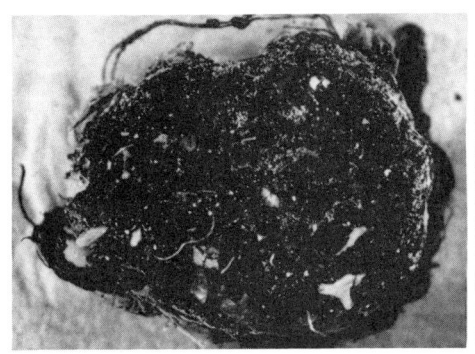

From beaches of the tropical Atlantic. Left: plastics embedded in a tarball; right: plastics in beach wrack. (Courtesy R. Jude Wilber, Sea Education Association.)

any craft. Plastic waste must be disposed of ashore, hopefully at a recycling facility. For large ships, this is a substantial holding problem.

Waste and toxic chemicals common to the island region include the oxygen depleting effluents from distilleries and sugar industries, all generally untreated before reaching the coastal waters. Persistent chemicals, such as DDT and DDE, that may be used in agriculture are an additional problem. And, of course, toxic chemicals can be a concern relative to the groundwater supply.

There was a time when one site in the area was considered as a potential dump area for wastes from far afield. I recall discussions, a few decades in the past, when serious thought was given to burying toxic substances, including low-level radioactive wastes, in deep ocean trenches. The trench north of Puerto Rico seemed to be a natural for this. Such thinking lost out on three counts. First, successive upward migrations in the food chain from one step to another might bring toxics to the overlying waters. Second, we cannot be completely confident that there would be no physical dispersal out of a trench. And, finally, while superficially a trench may seem like a dead chasm, it is in reality a geologically "restless" site where tectonic plates are interacting.

The incident I just mentioned where trash was carried out to be dispersed by the currents, points to the fact that on the smallest islands, finding a suitable site for a landfill is not easy. This, of course, does not excuse sheer carelessness in leaving rusting cars, piles of cans, etcetera, scattered in the yards, fields and even on the beautiful beaches.

For a discussion of regional efforts regarding pollution see comments in the write-up of "Law of the Sea" referring to the Cartagena Convention.

Observing and Photographing

Observing and photographing go hand in hand and both take a large measure of patience. There are three approaches to consider whether for observing, photographing, or both:

Remain quiet at a promising site until "nature becomes accustomed to your presence."

Move slowly and quietly, listening as well as observing, and pausing when something special is noted. I prefer this but such pauses are no substitute for more patient observing.

Cover more territory as you vigorously hike or swim. The exercise is good and you can enjoy the surroundings, but you are bound to miss important details.

As to photography underwater I'll concentrate on the uniqueness of the aquatic environment as it affects one's efforts. Generally, for best results you must be "comfortable" with and figure on using SCUBA. Even though bubbles from the regulator may disturb the surroundings, using diving gear is the only hope of working as patiently as is necessary. Admittedly this is contrary to my comments on how well you might get along by snorkeling and, being stubborn, I note that I have made some satisfying shots when snorkeling.

Underwater the quality and quantity of light is substantially different from what one is accustomed to. To begin with, until the sun is high in the sky there is considerable reflection off the surface and a corresponding diminution of light below. In fact, there is appreciable loss, even at noon, *unless* the sun happens to be almost directly overhead. Also troublesome is the lens action of large waves, concentrating light beneath the crests and dispersing it beneath the troughs. Most important, however, is the differential absorption of the wave lengths of light as, progressively, the reds, yellows, and soon the greens of the spectrum are absorbed (see write-up of "The Surrounding Sea"). In the shallows with water acting as a filter and with some of the color lost, it is desirable to be as close to the subject as possible. With increasing depth, green colors will persist, but ultimately everything looks blue if there is any light at all. I recall that my own blood looked green from a cut I sustained at little more than 30 feet below the surface.

One can deal with the loss of color with depth either by using flash or a strobe light or by accepting the interesting effect of the color as is. Blue was the prevailing color in the Kodachrome original of the spiny lobster photographed in about 75 feet of water (see print accompanying the write-up of the "Spiny Lobster").

Incidentally, if you use a strobe or flash attached directly to your camera, you are likely to get a scattering of light reflected off suspended aggregates, particles and plankton that you may not have been aware of. To minimize this, position the flash away from the body of the camera, concentrating more on the subject than on the water in front of the lens.

From the experience of trying to reach out for something underwater you are undoubtedly aware that subjects are not as

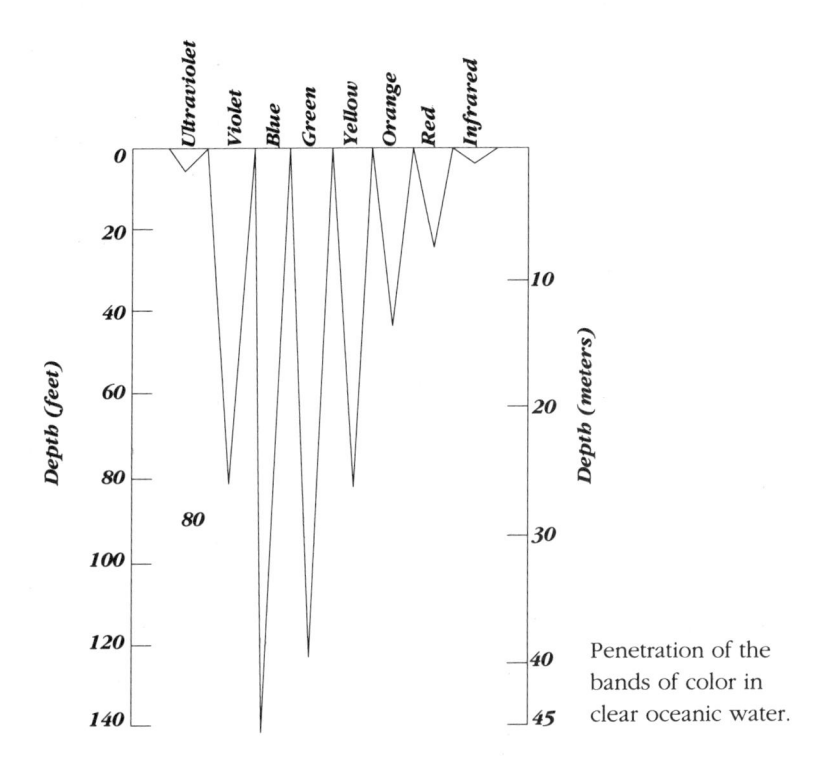

Penetration of the bands of color in clear oceanic water.

close as they seem. In other words, the distance as it appears to you is only about three quarters of the actual distance. This is due to the fact that the light rays come to a closer focal point as they bend at the face plate of your diving mask on passing from the water to the air medium. Since this refraction, as it is called, also occurs for light when it passes through the window of your camera, you should set the focus for the distance as you sense it rather than for any actual measurement. Of course, if you have reflex equipment, you are seeing the subject as it will register on the film. Domed ports, as widely available for underwater housings, correct for refraction, in other words, all light rays theoretically enter at a 90° angle and the camera and range finder are focussing in terms of the actual distance.

Some underwater cameras have built-in light meters. This is a decided advantage since the task of working underwater with a separate meter is very cumbersome. Without a meter I usually get along by applying the following rule of thumb. In the tropics, with reasonably clear water and with the sun well overhead, I get as close to the subject as possible and, using ASA 64 film, shoot at f8 and 1/125 seconds. Though this works for me, I hesitate to stress my satisfaction to anyone with greater expertise or proceding more thoroughly to obtain pictures of the highest quality possible. For improved color, someone has suggested that you open the lens one-half or even a whole stop more than any readings or your estimates suggest. Of course using flash overcomes the light limitations. One interesting suggestion, applicable in adequate light in shallow water, calls for aiming a flash on the primary subject with natural light filling in the background.

I also have one more rule of thumb: When conditions are favorable, about one out of every five shots taken underwater is usable according to my standards (not necessarily great, but usable). By comparison, I can use about half the pictures I take out of water.

This leads to a discussion of cameras and film. I will limit this to comments on conventional, still-picture cameras and elabora

Observing and Photographing 171

tions thereof, only mentioning in passing that today the possibilities in advanced techniques—motion photography, remote operations, and so on—are fantastic. If you are into such gear you are probably well beyond the basics discussed in this write-up.

For underwater photographing you may be equipped with or plan to purchase one of several good makes designed to operate to considerable depths. For simplicity while traveling I prefer a so-called amphibious camera that is handy for photography ashore and operates underwater without the need for an extra waterproof housing. A competent camera supplier can fill you in on exchangeable lenses, extension tubes, and other useful supplementary equipment.

Incidentally, customs officers often have a watchful eye for photographic equipment. If you take elaborate gear with you, carry a notarized list of this equipment, especially if it has a new look about it. If you buy considerable equipment overseas, as many do to take advantage of good purchases and/or duty-free sales, be prepared to pay a customs fee when you return.

As to film, don't count on buying an appreciable fresh supply in areas as remote as some of the islands we are discussing. A few passes through the airport security devices probably will do no harm to the film supply you take with you; however, since the x-ray effect is cumulative, repeated trips through the detectors may result in fogging. High speed films are the most sensitive. If, as a precaution, you wrap your film in lead foil, this will show in an x-ray detector (unless it is out of order). Since the inspector may want to check on such a package, have it where it can be readily taken out of your luggage. My contention is that you have a right to have your loaded camera handled and inspected separately.* Once an inspector insisted on shooting one exposure with my camera to make certain it was truly a camera. Seemed reasonable, especially since there was a terrorist threat at the time.

Finally, the light underwater in the tropics is almost always adequate for film of medium or slower speeds (i.e., ASA 25 or 64). I would argue for transparencies, since good prints can be made from these but you can't go from prints to transparencies. Whether using natural or artificial light, film such as Kodachrome with good rendition of reds and greens is widely recommended. On the other hand, there is something to be said for blue-sensitive film like Ektachrome when working in this environment where blue prevails.

Usually figure on having a freshly loaded camera when you begin a dive. You can't reload underwater.

My concluding comment is offered *with feeling*. Always be sure your equipment is firmly secured to your person. Toppled by a high wave across a reef in Fiji, I lost a camera that had seemed adequately secure with a strong strap around my neck. It was a Calypso made in France before the model was taken over and produced by Nikonos. The Calypso consistently caught the eyes of colleagues who had never seen this precedent-setting model. Somehow it always treated me better than the Nikonos I purchased to replace it.

*Unfortunately there are a few countries where this right is not respected. I challenge reluctant inspectors with the threat of reporting them by name and identification number.

Beware

In the Water

Portuguese Man-O-War

Contact with the long, trailing stingers can cause severe pain, nausea, and even difficulty in breathing. Stinging cells may survive and inflict pain even after the man-o-wars have been washed up on the beach.

Remove adhering material, using a stick, a knife or something similar, to avoid further contact.

Apply alcohol, dilute ammonia water, or soap and water, or methylated calamine lotion. Talcum or baking powder may help relieve pain. It is usually wise to seek medical help.

Sea Wasps (Cubomedusae)

Though a member of this group found along the coast of Australia is considered one of the most dangerous marine animals, those in the Caribbean are less virulent and generally only occur near the surface at night. The fact that they travel in schools compounds the distress they may cause.

Treatment as for man-o-war.

I hesitate to mention sea wasps since encounters are so unlikely; however, if there are any indications of their presence, wear protective clothing and, when diving, be sure there are none on the mouthpiece when switching from snorkel to regulator, or vice-versa.

Jellyfish (alias Sea Nettles)

Stings from contact are basically the same as from the Portuguese man-o-war, but far less severe.

Treatment is the same as for Portuguese man-o-war but pain is minor and subsides fast, unless the irritation is from the rare sea wasps mentioned above.

Millepore Corals

Often dubbed "fire" or "stinging corals" since contact causes pain not unlike that of jellyfishes and should be similarly dealt with. (The polyps of fire corals, being very small, are embedded in tiny pinholes in the relatively smooth skeletal surface.)

Coral Cuts

Most any foreign protein may aggravate an open sore. Since the slime from corals is definitely foreign, cuts from corals can become infected and be slow-healing.

Wash thoroughly with soap and water; apply hydrogen peroxide and alcohol; a multi-antibiotic ointment may be helpful.

Generally the non-millepore corals and, with a few exceptions, the anemones do not sting in the same sense as the three preceding groups, even though they have the same basic stinging cells which are generally used to stun prey.

Fire and Touch-Me-Not Sponges

Contact may result in itching, swelling and pain.

Actually many sponges have siliceous (glassy) spicules that

can be very irritating on contact, but apparently the spicules on these two sponges not only irritate but transmit a poison. Wearing gloves for handling any sponge is probably a good precaution.

Wash thoroughly; apply alcohol or dilute ammonia water.

Various Bristle Worms

The bristles are also siliceous spicules.

Since these bristles accumulate on the surface of the skin, initial rubbing or washing may only aggravate the soreness. First try to remove them with the sticky side of a tape.

Treat as for similar irritations from sponges.

Long-spined and other Sea Urchins

The poisonous spines break off under the skin causing pain and swelling that may last for several hours.

Removing fragments begins to take on the aura of minor surgery but, if you can dislodge some bits and pieces, you are a bit ahead.

Treat with dilute ammonia water and an external antibiotic; also just wait it out. I had an encounter off Jamaica, getting a number of spines in the ankle. Local fishermen nearby shouted earthy advice: "Piss on it." This really is a tried and true technique.

Stingrays and Scorpionfishes

There is always an outside chance of stepping on the poisonous spines of one of these semi-buried, relatively sedentary bottom dwellers.

With the spine or spines removed, let bleed profusely, applying a tourniquet if necessary; soak in very hot or very cold water; apply an external antibiotic.

You may need medical attention.

Spines in General

Most will at least cause a foreign protein reaction and some are actually poisonous.

Best general suggestion—treat with dilute ammonia water and an external antibiotic.

Skin Reactions

Surface slimes from various critters, including sea cucumbers and sea slugs, may cause mild skin reactions.

Wash thoroughly; apply alcohol.

Octopus Bites

The bite transmits an anticoagulant.

Gauze and pressure may be needed to stop bleeding; use an external antibiotic.

Sharks

Yes, they can be dangerous but, with a little common sense, the threat is minimal.

Stay out, or get out of the water if you know sharks are around. If you must spearfish, don't swim with bleeding carcasses trailing behind. Sometimes sharks aggregate in feeding frenzies brought on by an abundance of what they regard as "goodies."—Stay Clear.

During a "swim call" it is wise to post a shark watch. It is just as well not to swim at night unless you are very sure of your surroundings.

Tar

It's just a *menace*. Remove with turpentine or mineral spirits. I would be a bit cautious in using cleaning solutions offered for the convenience of bathers. If I were to use any of these regularly, I would inquire as to the chemical content and seek responsible opinion as to possible effects.

Danger Likely to be Exaggerated

Barracudas—Vicious looking but "tame". Don't corner them.

Moray Eels—Their bite can be vicious, but they generally stay in their holes, at least during the day. I assume you are not about to put your hand in front of a moray to be bitten. If you are a victim, clean well and apply an external antibiotic.

Cone Shells—Some are deadly poisonous when extended and "biting" but the worst ones are elsewhere, mostly in the Pacific. Even so, caution is recommended.

Water Safety

I will not elaborate on the usual rules of safety while swimming and the need for thorough training if using SCUBA, but I will add that you should anticipate the possibility of dangerous currents. While the current behind and even near the crest of a reef will usually carry you safely toward shore, the opposite is possible, particularly through passes releasing water that has entered a lagoon across a reef crest. If you are swept seaward from the shore, bear in mind that you can soon surface. With your wits about you, take note of the fact that your direction of drift is likely to be oblique to the shoreline, particularly if you have been carried out by an undertow. Don't swim against, but at right angles to this drift. Finally, bear in mind that in the water you are mighty inconspicuous to passing boaters some of whom are wild operators. In your exploring you will often be outside the safety areas where there are adequate lifeguards, so be especially careful. If diving with SCUBA make certain the appropriate warning flags are conspicuously displayed.

On Land

Mosquitoes

They can be a nuisance but there's nothing to worry about. Yellow fever, severe in the past, is no longer a concern. The same is true for malaria, and dengue fever is rare. Use repellent to counter the nuisance.

Sandflies or No-See-Ums

Annoying? Very! But health-wise harmless. Frequently mosquitoes and sandflies are worse during twilight hours than in total darkness. Use repellent.

Scorpions, Tarantulas, Centipedes

Can inflict poisonous stings. These pests are seldom in the open and stings are very unlikely. Generally they are not deadly.

Encourage profuse bleeding; apply an external antibiotic.

Seek medical attention if symptoms seem severe.

Fer-de-Lance

This is a very poisonous snake found on a few islands, most notably Saint Lucia and Martinique.

If bitten, let bleed profusely; if available use anti-venom according to directions; apply a tourniquet as needed and remain as quiet as possible.

Get medical attention immediately. Hopefully the necessary anti-venom will be on hand, or can be readily obtained where you show up for treatment.

Bees and Wasps

Stings are troublesome and sometimes temporarily painful. Not serious unless you have a unique sensitivity. Hopefully, those with such a sensitivity know of the necessary antidote to meet their specific needs and carry this with them.

Manchineel

Sap from leaves, twigs and bark is very irritating.

Wash thoroughly; apply alcohol.

By all means *do not* eat the fruit which look like miniature green apples.

Plants in General

We can't cover everything. Common sense, such as staying clear of cactus spines and various nettles, should prevail. See comment below under "Local Knowledge."

Sun Exposure

The possibility of developing skin cancers from the ultraviolet rays of the sun is very real and is becoming an increasing problem with the depletion of ozone in the upper atmosphere. Superficial skin cancers, not serious if treated early, can lead to malignant cancerous growths. Get an effective sun block (one to which you are not allergic) and apply generously and regularly. Cover up as much as possible and avoid excessive sun exposure, especially when the sun is high overhead. (Unfortunately this is the best time for underwater viewing).

    P.S. Don't brag about your deep tan.—It's a sign of neglect.

Drinking Water

If in doubt, boil before using. Also thoroughly wash vegetables and fruit. Schistosomiasis (bilharzia) was once common in St. Lucia, but is now negligible due to an eradication program.

Local Knowledge

I rely heavily on local knowledge. If the natives report a common danger or poison, I respect their warning. For me, a check with the locals is very reassuring.

Disclaimers

I am deliberately avoiding a review of toxic foods. The subject is too broad, ranging from specific toxins, to spoilage, to preparation practices, and even allergies. You should at least be aware, however, of the poison known as ciguatera, causing muscular weakness, numbness, temperature reversal sensations, etcetera. The basic cause is apparently a toxin in certain one-celled algae sometimes occurring around reefs. Through the food chain this reaches toxic levels in the flesh of higher carnivores. Seek local knowledge. Also—don't include any of the puffers or porcupine fishes in your diet. The poisons in their flesh is not ciguatera but the toxic effects are similar. Medical attention is important for both ciguatera and puffer poisoning.

Beware 181

While there are many interesting ways in which natives use wild fruits, leaves, roots, and so on, you must use extreme caution. Don't be an enthusiastic but uninformed amateur.

I hesitated to write a section titled "Beware" for four reasons:

One can never be sure that all possibilities are covered. In fact stings, rashes, and so forth can occur and remain a mystery.

Sensitivities vary, and we can't rule out some extreme individual reactions to toxins that are generally mild.

Adequate first aid is a subject in itself, beyond the scope of this book. Diving and swimming accidents and their causes merit special attention.

When all possibilities are listed the total can seem ominous, giving an exaggerated sense of danger. Risks on land are almost negligible (except for some wild drivers) and, as to the very inviting underwater scene, the greatest danger is man's limited ability to function in the water.

Appendices

Introduction

Using the Checklists

A common first reaction to the rich and varied display of life in the tropics, particularly when snorkeling over a coral reef, is to be thrilled but perhaps a bit dazzled. Trying to tell just what you are seeing is not as difficult as it may seem if, with suitable field guides (see next section), you start with the more obvious and sort things out one step at a time. For good observing it helps to keep a log on suitable checklists. There are eight such appendices to this book, namely:

Fishes—Fairly complete, though for many groups the listing of separate species had to be limited due to the difficulty of separating them in the field. Some species that only occur at appreciable depths are omitted. A complete list of the Sea Turtles is entered at the end of this list of Fishes.

Corals and Gorgonians—Relatively complete for stony corals; less so for anemones and soft corals where many species are lumped.

Shells—Emphasis is on the larger forms and those most readily obtained without dredging. Many very small species, i.e. less than ¼ inch, are not included.

Marine Invertebrates—Emphasis is on unique and common inverebrates that are found quite readily and identified.

Marine Bottom Plants—Limited selection as for marine invertebrates.

Birds—Relatively complete list catering to the interests of avid birders.

Flowers, Shrubs and Vines—Emphasis is on the highlights, especially the plants that are showy, unique, generally abundant, or otherwise interesting.

186   *Introduction*

Trees—Limited listing as for flowers, etc.

Microscopic life is not listed. Thinking in terms of the more usual interests of travelers, life in freshwater and the vertebrates on land are skipped. Insects, though very important ecologically, are also by-passed.

In general families or orders are listed first, using a common group name where widely recognized, with the scientific family name following in parenthesis. Some of the species you are likely to see are written below and a few blanks are added in case you encounter and identify others. The practice followed is to give a common name followed by the genus and species in proper scientific form, i.e. in italics and in lower case type except for the first letter of the genus.

It is useful here to explain the meaning of "sp." and "spp." as these abbreviations sometimes appear after the genus name. Both indicate recognition of the given genus. "Sp." provides for indicating the genus, yet allows for uncertainty as to what species is involved. This may even be a new species, thus far undescribed and yet to be named by taxonomists. "Spp." also indicates recognition of the genus, but expresses the opinion that there is more than one species to be identified.

The most obvious use of the lists is for individuals to check off their own; however, a team effort might be rewarding for persons traveling as a group. Try some competition, with individuals initialing entries as fast as they make observations. Incidentally, you may be able to identify a family though unable to single out any species. The blank following the family name can be checked in such cases. Hobbyists and specialists, such as birders and shell collectors, are often a great help.

The field guides suggested include many interesting notes on the species listed. Reading these enriches one's nature experience.

Finally, you never know when you will want a little magnification. I tend to carry a small hand lens and compact pocket-size field glasses.

Use Discretion, Don't Plunder

It is O.K. to take empty shells from the beach—though this may leave little for the next observer. Except for parks and gardens, it is fair enough to pluck an occasional flower where they are blooming in profusion. On the other hand, taking anything alive off a reef, especially the corals, sea fans, and live shellfish, destroys reef structure. —Besides, nature doesn't look all that great back home when dried out and faded. *The "camera bug" is the greater nature lover.*

Some Useful Field Guides

You might load yourself down with field guides and, even then, wish you had more. But don't be discouraged. Start out with two or three guides selected to meet your primary interests. If you have a further need, more field guides can usually be found in local shops in the Caribbean.

I. Starting with marine life *I note the following guide books:*

Guide to Corals and Fishes of Florida, the Bahamas and the Caribbean. 1977. Author: Idaz Greenburg. Waterproof edition available with Jerry Greenberg as co-author.

or

The Living Reef, Corals and Fishes of the Tropical Atlantic. 1972. Authors: Jerry and Idaz Greenberg. Both books published by Seahawk Press, 6840 S.W. 92nd St., Miami, FL 33156. All pictures for the two references above are in color, including many fishes, common corals and gorgonians, all sea turtles and miscellaneous other entries.

Caribbean Seashells. 1962. Authors: G. L. Warmke and R. T. Abbott. Dover Publications, 31 East Second Street, Mineola, NY 11501.

For additional illustrations you might wish to supplement this with:

American Seashells. 1974. Author: R. Tucker Abbott. Van Nostrand Rheinhold Co., 115 Fifth Ave., NY 10003 (no longer available directly from the publisher).

Marine Plants of the Caribbean—A Field Guide from Florida to Brazil. 1989. Authors: D. S. Littler, M. M. Littler, K. E. Bucher and J. N. Norris. Smithsonian Institution Press, 470 L'Enfante Plaza, Washington, DC 20560. Comprehensive and with good color plates.

Seashore Life of Florida and the Caribbean. 1980. Author: Gilbert Voss. Banyan Books, P.O. Box 431160, Miami, FL 33143. Comprehensive

Some Useful Field Guides

coverage of marine plants and invertebrates (except the shellfish or molluscs, omitted since that subject is covered by other field guides). Good illustrations in black and white, some representative color photos.

II. For observations on the spectacular plants of the islands you may consider:

Tropical Blossoms of the Caribbean. 1960. Authors: Dorothy and Bob Hargreaves. Ross-Hargreaves Publishers, Box 11897, Lahaina, HI 96761.

Tropical Trees [Latin American Countries]. 1965. Authors: Dorothy and Bob Hargreaves. Publishers as above.

200 Conspicuous, Unusual, or Economically Important Tropical Plants of the Caribbean. 1988. John M. Kingsbury. Bullbrier Press, 10 Snyder Heights, Ithaca, NY 14850.

Flowers of the Caribbean. 1978. Authors: G. W. Lennox and S. A. Seddon. Macmillan Caribbean, Houndmills Basingstoke Hampshire, RG21 2XS, England.

Trees of the Caribbean. 1980. Authors: S. A. Seddon and G. W. Lennox. Macmillan Caribbean (see address above).

All of the above have color photos. Bear in mind that such guides to the flora only touch upon the more spectacular and beautiful and that a full treatment requires a number of comprehensive treatises as used by specialists.

III. And for birds I mention:

Birds of the Eastern Caribbean. 1990. Author: Peter Evans. Macmillan Caribbean (see address above).

Birds of the West Indies. 1985. Author: James Bond. Houghton Mifflin Co., Order Processing, Wayside Road, Burlington, MA 01803.

A Guide to the Birds of Puerto Rico and the Virgin Islands. 1989. Author: Herbert A. Raffaele. Princeton University Press, Ordering Department, 3175 Lawrenceville Pike, Lawrenceville, NJ 08648. This book has better illustrations than the guide by Bond.

Some Useful Field Guides

IV. More useful guides. Some readers may be partial to the following very good guides:

Handguide to Coral Reef Fishes of the Caribbean. 1980. Author: F. Joseph Stokes. Lippincott and Crowell, East Washington Sq., Philadelphia, PA 19105. Comprehensive guide with good illustrations.

Fishes of the Caribbean Reefs, the Bahamas and Bermuda. 1978. Ian F. Took. Macmillan Caribbean (see address above). Good color photographs of a number of the fishes.

Fishwatchers Guide to West Atlantic Coral Reefs. 1972. Author: C. G. Chapin and Peter Scott. Harrowood Books, 3943 N. Providence Road, Newton Square, PA 19073. Good color illustrations. Waterproof edition available.

Caribbean Wild Plants and Their Uses. 1986. Author: Penelope N. Honychurch. Macmillan Caribbean (see address above). Comprehensive listings. Good line drawings.

Seashell Treasures of the Caribbean. 1986. Author: Lesley Sutty, edited by R. Tucker Abbott. Macmillan Caribbean (see address above) (no longer available directly from the publisher). Not a comprehensive treatise in the manner of *Caribbean Seashells* by Warmke and Abbott yet, with its selected illustrations, this has considerable appeal.

Collectible Shells of Southeastern United States, Bahamas and Caribbean. 1984. R. Tucker Abbott. American Malacologists Inc., P.O. Box 2255, Melbourne, FL 32902. All color photos.

A Field Guide to Coral Reefs: Caribbean and Florida. 1982. Author: Eugene H. Kaplan. Houghton Mifflin Co. (see address above). Comprehensiveness results in shortcomings in dealing with some groups including Fishes and Shells.

Caribbean Reef Invertebrates and Plants. 1978. Author: Patrick I. Colin. T. F. H. Publications, One T. F. H. Plaza, Third and Union Avenues, Neptune, NJ 07753. Title changed to *Marine Invertebrates and Plants of the Living Reefs,* without change in content. Extensive discussion of the biota covered. Many colored photos. Not exhaustive for some of the groups dealt with.

Fruits and Vegetables of the Caribbean. 1988. Authors: M. J. Bourne, G. W. Lennox and S. A. Seddon. Macmillan Caribbean (see address above). Good color illustrations.

IV. Finally, those wishing a more definitive treatise on fishes *of the reef might wish to refer to:*

Caribbean Reef Fishes. 1968. Author: John E. Randall. T. F. H. Publications (see address above).

V. For some highlights, all too brief, on insects *the following is widely available in bookstores and other shops:*

Butterflies and Other Insects of the Eastern Caribbean. 1986. Author: P. D. Stiling, Macmillan Caribbean (see address above).

If you are frustrated by so many suggestions you may appreciate some thoughts on getting along with a minimum. One might choose *The Living Reef* by Jerry and Idaz Greenberg. The rest of your needs might be met by selecting according to your special interests. For starts, many of the islands have botanical gardens with labels on the more interesting higher plants. As to shells and other invertebrates, you may be familiar already with some of the more common types, and you may have a chance to expand your knowledge with the help of traveling companions plus any books that happen to be handy where you are traveling. The same is true for the birds. And when it comes to marine plants, you can at least sort out most of the algae as greens, browns or reds.

Though waterproof guides are useful for wet working conditions, you may find them to be less useful for extensive underwater use.

Note: Macmillan Caribbean points out that its books are readily available in bookshops in the Caribbean. While not as available in the United States, the books may be obtained by writing their address in England. Some publishers gave an ordering address different from the publication location, in which case the place for ordering is used.

Fishes* (Plus Sea Turtles)

A challenging goal is to spot and recognize half the major groupings. While some are already familiar, seeing them in their native surroundings is a decided plus.

Major Groupings†

Sharks
Rays
Tarpons and
 Ladyfishes
Bonefishes
Herrings, Anchovies
 and Silversides
Lizard Fishes or
 Sand Divers
Moray, Conger and
 Snake Eels
Needlefishes
Halfbeaks
Trumpetfishes and
 Cornetfishes
Flyingfishes
Squirrel fishes

Sea Horses and
 Pipefishes
Barracudas
Mullets
Threadfins
Flounders and
 Tonguefishes
Cardinal fishes
Snooks
Groupers, Basses,
 Hinds and Hamlets
Fairy Basslets
Soapfishes
Bigeyes
Sweepers
Tilefishes
Cobias

Remoras
Jacks
Dolphins
Snappers
Grunts
Porgies
Croakers and Drums
Mojarras
Goatfishes
Sea Chubs
Spadefishes
Angelfishes and
 Butterflyfishes
Damselfishes
Hawkfishes
Wrasses and
 Hogfishes

*The word "fishes" is the correct plural when referring to more than one species. This may seem a little awkward at first.
†Some emcompass more than one family of fishes.

Fishes (Plus Sea Turtles)

Parrotfishes
Blennies
Tunas and
  Mackerel
Marlins and
  Sailfishes
Butterfishes
Gobies

Dragonets
Scorpionfishes
Flying Gurnards
Surgeonfishes
Triggerfishes
Filefishes
Trunkfishes and
  Cowfishes

Puffers
Porcupine Fishes
Clingfishes
Toadfishes
Frog fishes
Batfishes

Some of the identifications within the groups are difficult; others are relatively easy. Among the sharks, for example, the nurse shark is readily singled out, but many of the rest are far less distinct. Among the parrotfishes, species distinctions that may be difficult to begin with, are complicated by the variations that occur. Even the spectacular *blue* has color phases. And, when you feel satisfied that you have identified the beautiful queen angelfish, you may have only half the story; the young are quite different. —So it goes.

The five species of sea turtles that occur in the Eastern Caribbean are entered on the last page of this checklist.

Sharks (Several Families) ___
___ Nurse Shark, *Ginglymostoma cirratum*
___ Hammerhead Sharks, *Sphyrna* spp.
___ Tiger Shark, *Galeocerdo cuvieri*
___ Lemon Shark, *Negaprion brevirostrus*
___ Sharpnose Shark, *Rhizoprionodon porosus*
___ Requiem Sharks, *Carcharhinus* spp.
___ _____
___ _____

Rays (Four Families) ___
___ Southern Stingray, *Dasyatis americana*
___ Yellow Stingray, *Urolophus jamaicensis*
___ Spotted Eagle Ray, *Aetobatis narinari*
___ Atlantic Manta Ray, *Manta birostris*
___ _____

Fishes (Plus Sea Turtles)

Tarpons and Ladyfishes (Elopidae) ___
___ Ladyfish, *Elops saurus*
___ Tarpon, *Megalops atlanticus*

Bonefishes (Albulidae) ___
___ Bonefish, *Albula vulpes*

Herrings, Anchovies and Silversides (Three Families) ___ Silver-sided, often schooling, and mostly small, plankton-feeding fishes. Since they are so difficult to distinguish underwater, no species differentiations are offered. However, in the marketplace you may be able to recognize the Anchovies as a group since the mouth opens almost to the gill cover. Also, the Thread Herring, *Opisthonema oglinum*, may be easy to single out as it is larger than the rest and has a long dorsal fin ray.

___ _____
___ _____
___ _____

Lizard Fishes or Sand Divers (Synodontidae) ___
___ Sand Diver, *Synodus intermedius*
___ _____

Spaghetti Eels (Moringuidae) ___
___ Spaghetti Eel, *Moringa edwardsi*
___ _____

Moray Eels (Muraenidae) ___
___ Spotted Moray, *Gymnothorax moringa*
___ Green Moray, *Gymnothorax funebris*
___ _____

Conger Eels (Congridae) ___
___ Garden Eel, *Heteroconger halis*
___ _____

Snake Eels (Ophichthidae) ___
___ Goldspotted Snake Eel, *Myrichthys oculatus*
___ _____

Needlefishes (Belonidae) ___
___ Redfin Needlefish, *Strongylura notata*
___ Houndfish, *Tylosurus crocodilus*
___ _____

Halfbeaks (Hemiramphidae) ___
___ Ballyhoo, *Hemiramphus brasiliensis*
___ _____

Trumpetfishes (Aulostomidae) ___
___ Trumpetfish, *Aulostomus maculatus*

Cornetfishes (Fistutariidae) ___
___ Bluespotted Cornetfish, *Fistularia tabacaria*

Flyingfishes (Exocoetidae) ___
___ Atlantic Flyingfish, *Cypselurus heterurus*

Fishes (Plus Sea Turtles)

Squirrelfishes and Soldierfishes (Holocentridae) ___
___ Longspined Squirrelfish, *Holocentrus rufus*
___ Reef Squirrelfish, *Holocentrus coruscus*
___ Dusky Squirrelfish, *Holocentrus vexillarius*
___ Blackbar Soldierfish, *Myripristis jacobus*
___ Cardinal Soldierfish, *Plectrypops retrospinus*
___ _____
___ _____

Pearlfishes (Carapidae) ___
___ Pearlfish, *Carapus bermudensis*

Seahorses and Pipefish (Syngnathidae) ___
___ Longsnout Sea Horse, *Hippocampus reidi*
___ Caribbean Pipefish, *Syngnathus rousseau*
___ _____
___ _____

Barracudas (Sphyraenidae) ___
___ Great Barracuda, *Sphyraena barracuda*
___ _____

Mullets (Mugilidae) ___
___ White Mullet, *Mugil curema*
___ _____

Threadfins (Polynemidae) ___
___ Barbu, *Polydactylus virginicus*
___ _____

Flounders and Tonguefishes (Three Families) ___
___ Peacock Flounder, *Bothus lunatus*
___ Caribbean Tonguefish, *Symphurus arawak*
___ _____
___ _____

Cardinal Fishes (Apogonidae) ___
___ Flamefish, *Apogon maculatus*
___ Freckled Cardinalfish, *Apogon conklini*
___ Belted Cardinalfish, *Apogon townsendi*
___ Conchfish, *Astrapogon stellatus*
___ _____
___ _____

Snooks (Centropomidae) ___
___ Snook, *Centropomus undecimalis*

Groupers and Sea Basses, including Hinds, Basslets and Hamlets (Serranidae) ___
___ Jewfish, *Epinephelus itajara*
___ Mutton Hamlet, *Alphestes afer*
___ Nassau Grouper, *Epinephelus striatus*
___ Rock Hind, *Epinephelus adscensionus*
___ Red Hind, *Epinephelus guttatus*

Fishes (Plus Sea Turtles)

(Serranidae cont.)
___ Graysby, *Cephalopholis cruentatum*
___ Coney, *Cephalopholis fulva*
___ Yellowfin Grouper, *Mycteroperca venenosa*
___ Tiger Grouper, *Mycteroperca tigris*
___ Black Grouper, *Mycteroperca bonaci*
___ Hamlets, *Hypoplectrus* spp.
___ Harlequin Bass, *Serranus tigrinus*
___ _____
___ _____

Fairy Basslets (Grammidae) ___
___ Fairy Basslet or Royal Gramma, *Gramma loreto*
___ _____

Soapfishes (Grammistadae) ___
___ Greater Soapfish, *Rypticus saponaceous*
___ _____

Bigeyes (Priacanthidae) ___
___ Glasseye Snapper, *Priacanthus cruentatus*
___ _____

Sweepers (Pempheridae) ___
___ Glassy Sweeper, *Pempheris schomburgki*
___ _____

Tilefishes (Branchiostegidae) ___
___ Sand Tilefish, *Malacanthus plumieri*
___ _____

Cobias (Rachycentridae) ___
___ Cobia, *Rachycentron canadum*

Remoras (Echeneidae) ___
___ Remora, *Remora remora*
___ Sharksucker, *Echeneis naucrates*
___ _____

Jacks (Carangidae) ___
___ Rainbow Runner, *Elagatis bipinnulatus*
___ Bar Jack, *Caranx ruber*
___ Blue Runner, *Caranx fusus*
___ Yellow Jack, *Caranx bartholomaei*
___ Crevalle Jack, *Caranx hippos*
___ Permit, *Trachinotus falcatus*
___ Palometa, *Trachinotus goodei*
___ African Pompano, *Alectis crinitus*
___ Greater Amberjack, *Seriola dumerili*
___ _____
___ _____

Dolphins (Coryphaenidae) ___
___ Dolphin, *Coryphaena hippurus*
___ _____

Snappers (Lutjanidae) ___
___ Gray Snapper, *Lutjanus griseus*
___ Schoolmaster, *Lutjanus apodus*
___ Mutton Snapper, *Lutjanus analis*

Fishes (Plus Sea Turtles)

(Lutjanidae cont.)
___ Mahogany Snapper, *Lutjanus mahogoni*
___ Lane Snapper, *Lutjanus synagris*
___ Dog Snapper, *Lutjanus jocu*
___ Other Snappers, *Lutjanus* spp.
___ Yellowtail snapper, *Ocyurus chrysurus*
___ _____
___ _____

Grunts (Pomadasyidae) ___
___ French Grunt, *Haemulon flavolineatum*
___ Bluestriped Grunt, *Haemulon sciurus*
___ Margate, *Haemulon album*
___ Smallmouth Grunt, *Haemulon chrysargyreum*
___ Tomtate, *Haemulon aurolineatum*
___ Other Grunts, *Haemulon* spp.
___ Porkfish, *Anisotremus virginicus*
___ Black Margate, *Anisotremus surinamensis*
___ _____
___ _____

Porgies (Sparidae) ___
___ Pluma, *Calamus pennatula*
___ Sea Bream, *Archosargus rhomboidalis*
___ _____

Croakers and Drums (Sciaenidae) ___
___ Reef Croaker, *Odontoscion dentex*

(Sciaenidae cont.)
___ Cubbyu or High-Hat, *Equetus acuminatus*
___ Spotted Drum, *Equetus punctatus*
___ _____

Mojarras (Gerreidae) ___
___ Yellowfin Mojarra, *Gerrus cinereus*
___ Mottled Mojarra, *Eucinostomus lefroyi*
___ Bigeye Mojarra, *Eucinostomus havana*
___ _____

Goatfishes (Mullidae) ___
___ Yellow Goatfish, *Mulloidichthys martinicus*
___ Spotted Goatfish, *Pseudupeneus maculatus*

Sea Chubs (Kyphosidae) ___
___ Bermuda Chub, *Kyphosus sectatrix*
___ _____

Spadefishes (Ephiddae) ___
___ Atlantic Spadefish, *Chaetodipterus faber*
___ _____

Angelfishes and Butterflyfishes (Chaetodontidae) ___
___ Rock Beauty, *Holocanthus tricolor*
___ Queen Angelfish, *Holocanthus ciliaris*

Fishes (Plus Sea Turtles)

(Chaetodontidae cont.)
___ Blue Angelfish, *Holocanthus isabelita*
___ Gray Angelfish, *Pomacanthus arcuatus*
___ French Angelfish, *Pomacanthus paru*
___ Spotfin Butterflyfish, *Chaetodon ocellatus*
___ Foureye Butterflyfish, *Chaetodon capistratus*
___ Banded Butterflyfish, *Chaetodon striatus*
___ Reef Butterflyfish, *Chaetodon sedentarius*
___ Longsnout Butterflyfish, *Chaetodon aculeatus*
___ _____
___ _____

Damselfishes (Pomacentridae) ___
___ Yellowtail Damselfish, *Microspathodon chrysurus*
___ Threespot Damselfish, *Stegastes planifrons*
___ Bicolor Damselfish, *Stegastes partitus*
___ Dusky Damselfish, *Stegastes fuscus*
___ Beaugregory, *Stegastes leucostictus*
___ Cocoa Damselfish, *Stegastes variabilis*
___ Sergeant Major, *Abedufduf saxatilis*
___ Brown Chromis, *Chromis multilineata*

(Pomacentridae cont.)
___ Blue Chromis, *Chromis cyanea*
___ _____
___ _____

Hawkfishes (Cirrhitidae) ___
___ Redspotted Hawkfish, *Amblycirrhitus pinos*

Wrasses and Hogfishes (Labridae) ___
___ Hogfish, *Lachnolaimus maximus*
___ Spanish Hogfish, *Bodianus rufus*
___ Bluehead, *Thalassoma bifasciatum*
___ Clown Wrasse, *Halichoeres maculipinna*
___ Slippery Dick, *Halichoeres bivittatus*
___ Puddingwife, *Halichoeres radiatus*
___ Yellowhead Wrasse, *Halichoeres garnoti*
___ Creole Wrasse, *Clepticus parrae*
___ Green Razorfish, *Hemipteronotus splendens*
___ _____
___ _____

Parrotfishes (Scaridae) ___
___ Queen Parrotfish, *Scarus vetula*
___ Princess Parrotfish, *Scarus taeniopterus*
___ Midnight Parrotfish, *Scarus coelestinus*
___ Striped Parrotfish, *Scarus iserti*
___ Blue Parrotfish, Scarus coeruleus

(Scaridae cont.)
___ Rainbow Parrotfish, *Scarus guacamaia*
___ Stoplight Parrotfish, *Sparisoma viride*
___ Redtail Parrotfish, *Sparisoma chrysopterum*
___ Bucktooth Parrotfish, *Sparisoma radians*
___ Redband Parrotfish, *Sparisoma aurofrenatum*
___ Yellowtail Parrotfish, *Sparisoma rubripinne*
___ _____
___ _____

Jawfishes (Opisthognathidae) ___
___ Yellowhead Jawfish, *Opisthognathus aurifrons*
___ _____

Blennies (Three Families) ___
___ Redlip (Hotlips) Blenny, *Ophioblennius atlanticus*
___ Sailfin Blenny, *Emblemaria pandionis*
___ Rosy Blenny, *Malacoctenus macropus*
___ _____
___ _____

Tunas and Mackerels (Scombridae) ___
___ King Mackerel, *Scomberomorus cavalla*
___ Cero, *Scomberomorus regalis*
___ Bigeye Tuna, *Thunnus obesus*

(Scombridae cont.)
___ Blackfin Tuna, *Thunnus atlanticus*
___ Yellowfin Tuna, *Thunnus albicares*
___ Atlantic Bonito, *Sarda sarda*
___ Skipjack Tuna, *Euthynnus pelamis*
___ Wahoo, *Acanthocybium solanderi*
___ _____

Marlins and Sailfishes (Istiophoridae) ___
___ Blue Marlin, *Makaira nigricans*
___ White Marlin, *Tetrapturus albidus*
___ Sailfish, *Istiophorus platypterus*
___ _____

Butterfishes (Stromateidae) ___
___ Man-o-War Fish, *Nomeus gronovii*
___ _____

Gobies (Gobiidae) ___
___ Greenband Goby, *Gobiosoma multifasciatum*
___ Ninelined Goby, *Ginsburgellus novemlineatus*
___ Bridled Goby, *Coryphopterus glaucofracnum*
___ Masked Goby, *Coryphopterus personatus*
___ Goldspot Goby, *Gnatholepis thompsoni*
___ _____
___ _____

Fishes (Plus Sea Turtles)

Dragonets (Callionymidae) ___
___ Lancer Dragonet, *Callionymus bairdi*
___ _____

Scorpionfishes (Scorpaenidae) ___
___ Spotted Scorpionfish, *Scorpaena plumieri*
___ Plumed Scorpionfish, *Scarpaena grandicornis*
___ Goosehead Scorpionfish, *Scorpaena bergi*
___ Reef Scorpionfish, *Scorpaenodes caribbaeus*
___ _____

Flying Gurnards (Dactylopteridae)
___ (Doesn't fly or glide)
___ Flying Gurnard, *Dactylopterus volitans*

Surgeonfishes (Acanthuridae) ___
___ Blue Tang, *Acanthurus coeruleus*
___ Doctorfish, *Acanthurus chirurgus*
___ Ocean Surgeon, *Acanthurus bahianus*
___ _____

Triggerfishes (Balistidae) ___
___ Queen Triggerfish, *Balistes vetula*
___ Gray Triggerfish, *Balistes capriscus*
___ _____

Filefishes (Monacanthidae) ___
___ Fringed Filefish, *Monocanthus ciliatus*
___ Slender Filefish, *Monocanthus tuckeri*
___ Scrawled Filefish, *Alutera scripta*
___ Whitespotted Filefish, *Cantherhines macrocerus*
___ Orangespotted Filefish, *Cantherhines pullus*
___ _____

Trunkfishes and Cowfishes (Ostraciidae) ___
___ Trunkfish, *Lactophrys trigonus*
___ Smooth Trunkfish, *Lactophrys triqueter*
___ Spotted Trunkfish, *Lactophrys bicaudalis*
___ Scrawled Cowfish, *Acanthostracion quadricornis*
___ _____

Puffers (Two Families) ___
___ Bandtail Puffer, *Sphaeroides spengleri*
___ Caribbean Puffer, *Sphaeroides greeleyi*
___ Sharpnose Puffer, *Canthigaster rostrata*
___ _____

Porcupinefishes (Diodontidae) ___
___ Porcupinefish, *Diodon hystrix*
___ Web Burrfish, *Chilomycterus antillarum*
___ _____

Fishes (Plus Sea Turtles)

Clingfishes (Gobiesocidae) ___
___ Hourglass Clingfish, *Tomicodon fasciatus*
___ Tadpole Clingfish, *Arcos macropthalmus*

___ _____

Toadfishes (Batrachoididae) ___
___ Sapo Bocon, *Amphichthys cryptocentrus*
___ Gulf Toadfish, *Opsanus beta*

___ _____

Frogfishes (Antennariidae) ___
___ Sargassumfish, *Histrio histrio*
___ Longlure Frogfish, *Antennarius multiocellatus*
___ Splitlure Frogfish, *Antennarius scaber*

___ _____

Batfishes (Ogcocephalidae) ___
___ Shortnose Batfish, *Ogcocephalus nasutus*

___ _____

Sea Turtles (Two Families) ___
___ Leatherback Turtle, *Dermochelys coriacea*
___ Green Turtle, *Chelonia mydas*
___ Hawksbill Turtle, *Eretmochelys imbricata*
___ Loggerhead Turtle, *Caretta caretta*
___ Olive Ridley Turtle, *Lepidochelys olivacea*

Corals, Gorgonians and Anemones

First look for the more conspicuous species in their most common forms. You can probably get to know ten or so very quickly. Later you may observe variations in their life forms, sometimes even spectacular variants as to color, and you may also begin to distinguish some of the less obvious species.

Stony Corals

Stinging or Fire Coral
(Milleporidae) ___
___ Leafy Stinging Coral, *Millepora complanata*
___ Square Stinging Coral, *Millepora squarrosa*
___ Encrusting Stinging Coral, *Millepora alcicornis*

Blushing Star Corals
(Astrocoeniidae) ___
___ Blushing Star Coral, *Stephanocoenia michelinii*
___ _____

Cactus Corals (Seriatoporidae) ___
___ Green Cactus Coral, *Madracis decactis*
___ Yellow Pencil Coral, *Madracis mirabilis*
___ _____

Elkhorns and Staghorn Corals
(Acroporidae) ___
___ Staghorn Coral, *Acropora cervicornis*
___ Elkhorn Coral, *Acropora palmata*
___ Fused Staghorn, *Acropora prolifera*
___ _____

Leaf Corals (Agaricidae) ___
___ Sheet Coral, *Agaricia lamarcki*
___ Leaf Coral (3 varieties: plain, purple and massive), *Agaricia agaricites*
___ Ribbon Coral, *Agaricia tenuifolia*
___ Fragile Saucer Coral, *Agaricia fragilis*
___ Scroll Coral, *Agaricia undata*
___ Saucer coral, *Helioseris cucullata*
___ _____

Starlet Corals (Siderastreidae) ___
___ Rough Starlet Coral, *Siderastrea radians*

Corals, Gorgonians and Anemones

(Siderastreidae cont.)
___ Smooth Starlet Coral, *Sidterastrea siderea*
___ _____

Porous Corals (Poritidae) ___
___ Club Finger Coral, *Porites porites*
___ Finger Coral, *Porites furcata*
___ Mustard Hill Coral, *Porites astreoides*
___ Small Finger Coral, *Porites divaricata*
___ _____

Brain Corals (Faviidae) ___
___ Grooved Brain Coral, *Diploria labyrinthiformis*
___ Knobby Brain Coral, *Diploria clivosa*
___ Common Brain coral, *Diploria strigosa*
___ Giant Brain Coral, *Colpophyllia natans*
___ Cavernous Star Coral, *Montastrea cavernosa*
___ Mountainous Star Coral, *Montastrea annularis*
___ Lobed Star Coral, *Solenastrea hyades*
___ Smooth Star Coral, *Solenastrea bournoni*
___ Rose Coral, *Manicina areolata*
___ Golfball Coral, *Favia fragum*
___ Ivory Tube Coral, *Cladocora arbuscula*
___ _____

Bush Corals (Oculinidae) ___
___ Ivory Bush Coral, *Oculina diffusa*
___ Ivory Tree Coral, *Oculina valenciennes*
___ _____

_____(Meandrinidae) ___
___ Butterprint Brain Coral, *Meandrina meandrites*
___ Elliptical Star Coral, *Dichocoenia stokesii*
___ Pillar Coral, *Dendrogyra cylindrus*
___ _____

Fungus Corals (Mussidae) ___
___ Large Flower coral, *Mussa angulosa*
___ Fungus Coral, *Mycetophyllia lamarckiana*
___ Thin Fungus Coral, *Mycetophyllia aliciae*
___ Grooved Fungus Coral, *Mycetophyllia ferox*
___ Large-cupped Fungus Coral, *Scolymia lacera*
___ Sinuous Cactus Coral, *Isophyllia sinuosa*
___ Rough Star Coral, *Isophyllastrea rigida*
___ _____

Flower Coral (Caryophyliidae) ___
___ Flower Coral, *Eusmilia fastigiata*

Corals, Gorgonians and Anemones

Tube Corals (Dendrophylliidae) ___
___ Orange Tube Coral, *Tubastrea coccinea*

___ _____

Gorgonians, Anemones, Etc.

Sea Fingers (Briarareidae) ___
___ Corky Sea Fingers, *Briareum asbestinum*
___ Encrusting Brown Gorgonian, *Erythropodium caribaeorum*

___ _____

Sea Rods (Plexauridae) ___
___ Black Sea Rod, *Plexaura homomalla*
___ Common Sea Rod, *Plexaura flexuosa*
___ Spiny Candelabrum, *Muricea muricata*
___ Double-forked Plexaurella, *Plexaurella dichotoma*
___ Gray Plexaurella, *Plexaurella grisea*
___ Eunices, *Eunicea* spp.

___ _____
___ _____

Sea Feathers, Sea Fans, etc. (Gorgoniidae) ___
___ Forked Sea Feather, *Pseudopterogorgia bipinnata*
___ Slimy Sea Feather, *Pseudopterogorgia americana*
___ Smooth Sea Feather, *Pseudopterogorgia acerosa*
___ Other Sea Feathers and Sea Plumes, *Pseudopterogorgia* spp.

___ Common Sea Fan, *Gorgonia ventalina*
___ Venus Sea Fan, *Gorgonia flabellum*
___ Sea Whips, *Pterogorgia* spp.

___ _____

Sea Pansies (Renillidae) ___
___ Common Sea Pansy, *Renilla reniformis*

___ _____

Sea Anemones (Several Families) ___
___ Pink-tipped Anemone, *Condylactis gigantea*
___ Tricolor Anemone, *Calliactis tricolor*
___ Pale Anemone, *Aiptasia pallida*
___ Sea Mat, *Ricordia florida*
___ Ringed Anemone, *Bartholomea annulata*
___ Sun Anemone, *Stoichaetic helianthus*
___ Red Warty Anemone, *Bunodosoma granulifera*

___ _____
___ _____
___ _____

Zoanthids (Zoanthidae) ___
___ Green Colonial Anemone, *Zoanthus sociatus*
___ Knobby Zooanthid, *Palythoa caribaea*

___ _____
___ _____

Shells

(For related molluscs without shells, see Nudibranchs, Octopuses and Squids under "Marine Invertebrates")

Beachcombing for shells is fun but you can do more. Turn over rocks; collect at night; and while snorkeling, be on the lookout, not only on the bottom, but look for shells attached to plants and animals. When collecting, always restore the habitat as best possible.

You may already be familiar with several of the major groups of shells. Identifying species is not accomplished as readily. Variability is often the cause of difficult identifications but looking for variations can be both interesting and challenging. The common names used vary from place to place and even from one list to another. Furthermore, synonyms for scientific names are often encountered, even though only one is correct when the rules of taxonomy are properly adhered to. In the lists below I have followed the scientific names given in *Caribbean Seashells* by Warmke and Abbott who often mention synonyms in the descriptions given.

This list is a compromise, being limited to forms you may be most likely to find and distinguish, omitting tiny forms though they are noteworthy for ardent shellers, and omitting several deepwater forms and species you might miss unless you were dredging.

If there is a good shell collector in your ranks, this is a bonus. As in all such observing, a little expert help goes a long way.

Notes on preparing shells are given at the end of this checklist.

Chitons and Tusk Shells

Chitons (Several families) ___
___ Common West Indian Chiton, *Chiton tuberculatus*
___ Fuzzy Chiton, *Acanthopleura granulata*
___ _____
___ _____

Tusk Shells (Two Families) ___
___ Slender Cadulus, *Cadulus acus*
___ Ivory Tusk, *Dentalium eboreum*
___ _____
___ _____

Gastropods (Snails)

Limpets (Acmaeidae) ___
___ Antillean Limpet, *Acmaea antillarum*
___ _____

Keyhole Limpets (Fissurellidae) ___
___ Lister's Keyhole Limpet, *Diodora listeri*
___ Knobby Keyhole Limpet, *Fissurella nodosa*
___ Barbados Keyhole Limpet, *Fissurella barbadensis*
___ Rayed Keyhole Limpet, *Fissurella nimbosa*
___ _____

Top Shells and Tegulas (Trochidae) ___
___ West Indian Top Shell, *Cittarium pica*

(Trochidae cont.)
___ West Indian Tegula, *Tegula lividomaculata*
___ _____
___ _____

Turban and Star Shells (Turbinidae) ___
___ Chestnut Turban, *Turbo castanea*
___ Green Star Shell, *Astraea tuber*
___ Long-spined Star Shell, *Astraea phoebia*
___ Carved Star Shell, *Astraea caelata*
___ _____

Pheasant Shells (Phasianellidae) ___
___ Pheasant Shells, *Tricolia* spp.
___ _____

Nerites (Neritidae) ___
___ Bleeding Tooth, *Nerita peloronta*
___ Four-toothed Nerita, *Nerita versicolor*
___ Tesselated Nerita, *Nerita tesselata*
___ Zebra Nerita, *Puperita pupa*
___ Virgin Nerita, *Neritina virginea*
___ _____

Periwinkles (Littorinidae) ___
___ Zebra Periwinkle, *Littorina ziczac*
___ Angulate or Mangrove Periwinkle, *Littorina angulifera*
___ Common Prickly-winkle, *Nodilittorina tuberculata*

(Littorinidae cont.)
___ Beaded Periwinkle, *Tectarius muricatus*

_____

Truncatellas (Hydrobiidae) ___
___ Truncatellas, *Truncatella* spp.

_____

Turret Shells (Turritellidae) ___
___ Variegated Turret Shell, *Turritella variegata*
___ Knorr's Worm Shell, *Vermicularia knorri*

_____

Sundials (Architectonicidae) ___
___ Cylinder Sundial, *Heliacus cylindricus*
___ Common Sundial, *Architectonica nobilis*

_____

Worm Shells (Vermetidae) ___
___ Irregular Worm Shell, *Petaloconchus irregularis*
___ Decussate Worm Shell, *Serpulorbis decussata*

_____

Slit Worm Shells (Siliquariidae) ___
___ Slit Worm Shell, *Siliquaria anguillae*

_____

Caecums (Caecidae) ___
___ Caecums, *Caecum* spp.

_____

Planaxis Group (Planaxidae) ___
___ Planaxis, *Planaxis* spp.

_____

Modulus Group (Modulidae) ___
___ Atlantic Modulus, *Modulus modulus*

_____

Horn Shells (Potamadidae) ___
___ Turret Horn Shell, *Cerithidea costata*
___ False Cerith, *Batillaria minima*

_____

Ceriths (Cerithiidae) ___
___ Dwarf Cerith, *Cerithium variable*
___ Middle-spined Cerith, *Cerithium algicola*
___ Lettered Cerith, *Cerithium litteratum*
___ Ivory Cerith, *Cerithium eburneum*

_____

Triforas (Triphoridae) ___
___ Triforas, *Triphora* spp.

_____

Purple Sea Snails (Janthidae) ___
___ Purple Sea Snail, *Janthina janthina*

_____

Shells

Wentletraps (Epitoniidae) ___
___ Crenulated Wentletrap, *Opalia crenata*
___ West Atlantic Wentletrap, *Epitonium occidentale*
___ Lamellose Wentletrap, *Epitonium lamellosum*
___ _____

Melanellas (Eulimidae) ___
___ Cucumber Melanella, *Balcis intermedia*
___ _____

Hoof Shells (Hipponicidae) ___
___ False Cup-and-Saucer, *Cheilea equestris*
___ White Hoof Shell, *Hipponix antiquatus*
___ _____

Cup and Saucer and Slipper Shells (Calyptraeidae) ___
___ Spiny Slipper Shell, *Crepidula aculeata*
___ Convex Slipper Shell, *Crepidula convexa*
___ _____

Carrier Shells (Xenophoridae) ___
___ Atlantic Carrier Shell, *Xenophora conchyliophora*

Conchs (Strombidae) ___
___ Queen Conch, *Strombus gigas*
___ West Indian Fighting Conch, *Strombus pugilis*

(*Strombidae cont.*)
___ Hawk-wing Conch, *Strombus raninus*
___ Milk Conch, *Strombus costatus*
___ Rooster-tail Conch, *Strombus gallus*
___ _____

Trivias (Eratoidae) ___
___ Coffee Bean Trivia, *Trivia pediculus*
___ Suffuse Trivia, *Trivia suffusa*
___ _____

Cowries (Cypraeidae) ___
___ Measled Cowrie, *Cypraea zebra*
___ Atlantic Gray Cowrie, *Cypraea cinerea*
___ Atlantic Yellow Cowrie, *Cypraea spurca*
___ _____

Simnias and Flamingo Tongues (Ovulidae) ___
___ Simnias, *Neosimnia* spp.
___ Flamingo Tongue, *Cyphoma gibbosum*
___ _____

Moon Shells and Naticas (Naticidae) ___
___ Milk Moon Shell, *Polinices lacteus*
___ Brown Moon Shell, *Polinices hepaticus*
___ Morocco Natica, *Natica marochiensis*
___ Common Baby's Ear, *Sinum perspectivum*

Shells 209

(Naticidae cont.)
___ Colorful Atlantic Natica, *Natica canrena*

Bonnets and Helmets (Cassidae) ___
___ Atlantic Wood Louse, *Morum oniscus*
___ Scotch Bonnet, *Phalium granulatum*
___ King Helmet, *Cassis tuberosa*
___ Flame Helmet, *Cassis flammea*
___ Emperor Helmet, *Cassis madagascariensis*
___ Reticulated Cowrie-Helmet, *Cypraecassis testiculus*

Tritons (Ranellidae) ___
___ Trumpet Triton, *Charonia variegata*
___ Dog-head Triton, *Cymatium caribbaeum*
___ Gold-mouthed Triton, *Cymatium nicobaricum*
___ Atlantic Hairy Triton, *Cymatium pileare*
___ Knobby Triton, *Cymatium muricinum*
___ Angular Triton, *Cymatium femorale*
___ Atlantic Distorsio, *Distorsio clathrata*

Frog Shells (Bursidae) ___
___ Granular Frog Shell, *Bursa cubaniana*

Tuns (Tonnidae) ___
___ Atlantic Partridge Tun, *Tonna maculosa*

Rock Shells (Muricidae) ___
___ Apple Murex, *Murex pomum*
___ West Indian Murex, *Murex brevifrons*
___ Blackberry Drupe, *Drupa nodulosa*
___ Wide-mouthed Purpura, *Purpura patula*
___ Rustic Rock Shell, *Thais rustica*
___ Florida Rock Shell, *Thais haemastoma*
___ Deltoid Rock Shell, *Thais deltoides*
___ Two-sided Aspella, *Aspella paupercula*
___ Frilly Dwarf Triton, *Ocenebra intermedia*
___ Hexagonal Murex, *Muricopsis oxytatus*

Coral Shells (Magilidae) ___
___ Caribbean Coral Shell, *Coralliophila caribaea*
___ Short Coral Shell, *Coralliophila abbreviata*

Dove Shells (Columbellidae) ___
___ Common Dove Shell, *Columbella mercatoria*
___ Ovate Dove Shell, *Pyrene ovulata*
___ Beautiful Dove Shell, *Anachis pulchella*

Shells

(Columbellidae cont.)
___ White-spotted Dove Shell, *Nitidella ocellata*
___ Glossy Dove Shell, *Nitidella nitida*
___ Lunar Dove Shell, *Mitrella lunata*
___ _____
___ _____

Whelk Shells (Buccinidae) ___
___ Common Cantharus, *Cantharus auritulus*
___ Miniature Triton Trumpet, *Pisania pusio*
___ _____

Crown Conchs (Melongenidae) ___
___ West Indian Crown Conch, *Melongena melongena*

Mud Snails (Nassaridae) ___
___ Common Nassa, *Nassarius vibex*
___ _____

Tulips and Horse Conchs (Fasciolariidae) ___
___ True Tulip, *Fasciolaria tulipa*
___ Brown-lined Latirus, *Latirus infundibulum*
___ Short-tailed Latirus, *Latirus brevicaudatus*
___ Chestnut Latirus, *Leucozonia nassa*
___ White-spotted Latirus, *Leucozonia ocellata*
___ _____

Vase Shells (Turbinellidae) ___
___ Caribbean Vase, *Vasum muricatum*
___ _____

Olives (Olividae) ___
___ Netted Olive, *Oliva reticularis*
___ Caribbean Olive, *Oliva caribaeensis*
___ Caribbean Dwarf Olive, *Oliva petiolita*
___ _____

Miters (Mitridae) ___
___ Beaded Miter, *Mitra nodulosa*
___ Barbados Miter, *Mitra barbadensis*
___ _____

Volutes (Volutidae) ___
___ Music Volute, *Voluta musica*
___ _____

Nutmegs (Cancellariidae) ___
___ Common Nutmeg, *Cancellaria reticulata*
___ _____

Marginelids (Marginellidae) ___
___ Carmine Marginella, *Marginella haematita*
___ Orange-banded Marginella, *Hyalina avena*
___ _____

Cones (Conidae) ___
___ Crown Cone, *Conus regius*

(Conidae cont.)
___ Mouse Cone, *Conus mus*
___ Jasper Cone, *Conus jaspideus*
___ Julia's Cone, *Conus amphiurgus*
___ Glory of the Atlantic Cone, *Conus granulatus*

___ _____
___ _____

Augers (Terebridae) ___
___ Flame Auger, *Terebra taurinus*
___ Gray Auger, *Terebra cinerea*

___ _____

Turrets or Screw Shells (Turridae) ___
___ White Giant Turret, *Polystira albida*

___ _____
___ _____

Bubbles (Several Families) ___
___ Striated Bubble, *Bulla striata*
___ Elegant Paper Bubble, *Haminoea elegans*

___ _____
___ _____

Pyramid Shells (Pyramidellidae) ___
___ Brilliant Pyram, *Pyramidella candida*

___ _____
___ _____

Salt-marsh Snails (Ellobiidae) ___
___ Coffee Melampus, *Melampus coffeus*

___ _____

False Limpets (Two Families) ___
___ Goes' False Limpet, *Trimusculus goesi*

___ _____

Other Gastropods

___ _____
___ _____

Pelecypods (Bivalve Molluscs)

Nut Clams (Nuculidae) ___
___ Pointed Nut Clam, *Nuculana acuta*

___ _____

Ark Shells (Arcidae) ___
___ Turkey Wing, *Arca zebra*
___ Mossy Ark, *Arca imbricata*
___ White-bearded Ark, *Barbatia candida*
___ White Reticulated Ark, *Barbatia domingensis*
___ Red-brown Ark, *Barbatia cancellaria*
___ Doc Bales' Ark, *Barbatia tenera*
___ Adam's Miniature Ark, *Arcopsis adamsi*
___ Eared Ark, *Anadara notabilis*
___ Cut-ribbed Ark, *Anadara lienosa*
___ Blood Ark, *Anadara ovalis*
___ Incongruous Ark, *Anadara brasiliana*

___ _____

Shells

Bittersweets (Glycymerididae) ___
___ Comb Bittersweet, *Glycymeris pectinata*

___ _____

Mussels (Mytilidae) ___
___ Tulip Mussel, *Modiolus americanus*
___ Yellow Mussel, *Brachidontes citrinus*
___ Scorched Mussel, *Brachidontes exustus*
___ Hooked Mussel, *Brachidontes recurvus*
___ Black Date Mussel, *Lithophaga nigra*
___ Antillean Date Mussel, *Lithophaga antillarum*
___ Mahogany Date Mussel, *Lithophaga bisulcata*

___ _____
___ _____

Tree Oysters (Isognomonidae) ___
___ Flat Tree Oyster, *Isognomon alatus*
___ Lister's Tree Oyster, *Isognomon radiatus*

___ _____

Wing or Pearly Oysters (Pteriidae) ___
___ Atlantic Wing Oyster, *Pteria colymbus*
___ Atlantic Pearl Oyster, *Pinctada radiata*

Pen Shells (Pinnidae) ___
___ Amber Pen Shell, *Pinna carnea*
___ Spiny Pen Shell, *Atrina seminuda*

___ _____

Kitten's Paw (Plicatulidae) ___
___ Kitten's Paw, *Plicatula gibbosa*

___ _____

Scallops (Pectinidae) ___
___ Zigzag Scallop, *Pecten ziczac*
___ Ornate Scallop, *Chlamys ornata*
___ Lion's Paw, *Lyropecten nodosus*
___ Rough Scallop, *Aequipecten muscosus*
___ Calico Scallop, *Aequipecten gibbus*

___ _____
___ _____

Thorny Oysters (Spondylidae) ___
___ Atlantic Thorny Oyster, *Spondylus americanus*

Limas (Limidae) ___
___ Spiny Lima, *Lima lima*
___ Rough Lima, *Lima scabra*

___ _____

Jingle Shells (Anomiidae) ___
___ Common Jingle Shell, *Anomia simplex*
___ False Jingle Shell, *Pododesmus rudis*

Oysters (Ostreidae) ___
___ Coon Oyster, *Ostrea frons*
___ Caribbean or Mangrove Oyster,
   Crassostrea rhizophorae
___ _____

Carditas (Carditidae) ___
___ West Indian Cardita, *Cardita gracilis*

Coral Clam (Trapeziidae) ___
___ Coral Clam, *Coralliophaga coralliophaga*
___ _____

Diplodons (Diplodontidae) ___
___ Common Atlantic Diplodon, *Diplodonta punctata*
___ _____
___ _____

Lucines (Lucinidae) ___
___ Pennsylvania Lucina, *Lucina pensylvanica*
___ Thick Lucina, *Phacoides pectinatus*
___ Buttercup Lucina, *Anodontia alba*
___ Tiger Lucina, *Codakia orbicularis*
___ Cross-hatched Lucina, *Divaricella quadrisulcata*
___ _____
___ _____

Jewel Boxes (Chamidae) ___
___ Leafy Jewel Box, *Chama macerophylla*

(Chamidae cont.)
___ Little Corrugated Jewel Box, *Chama congregata*
___ Red Jewel Box, *Chama sarda*
___ Florida Jewel Box, *Chama florida*
___ Spiny Jewel Box, *Echinochama arcinella*
___ _____

Cockles (Cardiidae) ___
___ Yellow Cockle, *Trachycardium muricatum*
___ Prickly Cockle, *Trachycardium isocardia*
___ Spiny Paper Cockle, *Papyridea soleniformis*
___ Atlantic Strawberry Cockle, *Americardia media*
___ Common Egg Cockle, *Laevicardium laevigatum*
___ _____

Venus Clams (Veneridae) ___
___ Princess Venus, *Antigona listeri*
___ Rigid Venus, *Antigona rigida*
___ Cross-barred Venus, *Chione cancellata*
___ Beaded Venus, *Chione granulata*
___ White Pygmy Venus, *Chione pygmaea*
___ West Indian Pointed Venus, *Anomalocardia brasiliana*
___ Trigonal Tivela, *Tivela mactroides*
___ Lightning Venus, *Pitar fulminata*

(Veneridae cont.)
___ Purple Venus, *Pitar circinata*
___ Royal Comb Venus, *Pitar dione*
___ Calico Clam, *Macrocallista maculata*
___ Southern Dosinia, *Dosinia concentrica*
___ Atlantic Cyclinella, *Cyclinella tenuis*

___ _____
___ _____

Rock Borers (Petricolidae) ___
___ Boring Petricola, *Petricola lapicida*
___ Atlantic Rupellaria, *Rupellaria typica*

___ _____

Tellins and Macomas (Tellinidae) ___
___ Sunrise Tellin, *Tellina radiata*
___ Smooth Tellin, *Tellina laevigata*
___ Speckled Tellin, *Tellina listeri*
___ Alternate Tellin, *Tellina alternata*
___ Watermelon Tellin, *Tellina punicea*
___ Faust Tellin, *Arcopagia fausta*
___ Large Strigilla, *Strigilla carnaria*
___ Tagelus-like Macoma, *Macoma tageliformis*
___ Constricted Macoma, *Macoma constricta*
___ Atlantic Grooved Macoma, *Apolymetis intasriata*

___ _____
___ _____

Semeles and Abras (Semelidae) ___
___ White Atlantic Semele, *Semele proficua*
___ Purplish Semele, *Semele purpurascens*

___ _____
___ _____

Donax or Coquina (Donacidae) ___
___ Common Caribbean Donax, *Donax denticulatus*
___ Giant False Donax, *Iphigenia brasiliensis*

___ _____

Tagelus or Long-siphon Clams (Sanguinolariidae) ___
___ Gaudy Asaphis, *Asaphis deflorata*
___ Purplish Tagelus, *Tagelus divisus*
___ Stout Tagelus, *Tagelus plebeius*

___ _____

Razor Clams (Solenidae) ___
___ Antillian Jacknife Clam, *Solen obliquus*

___ _____

Surf Clams (Mactridae) ___
___ Winged Mactra, *Mactra alata*
___ Fragile Atlantic Mactra, *Mactra fragilis*

___ _____

Wedge Clams (Mesodesmatidae) ___
___ Common Ervilia, *Ervilia nitens*

___ _____

Corbulids (Corbulidae) ___
___ Corbulas, *Corbula* spp.

___ _____

Angel Wings (Pholadidae) ___
___ Angel Wing, *Cyrtopleura costata*
___ Wedge-Shaped Martesia, *Martesia cuneiformis*
___ Striate Martesia, *Martesia striata*

___ _____

Shipworms (Teredinidae) ___
___ Teredos, *Teredo* spp.
___ Bankias, *Bankia* spp.

___ _____

Paper Shells (Lyonsiidae) ___
___ Pearly Lyonsia, *Lyonsia beana*

___ _____

Pandoras (Pandoridae) ___
___ Southern Pandora, *Pandora bushiana*

___ _____

Dipper Shells (Cuspidariidae) ___
___ West Indian Cuspidaria, *Cardiomya perrostrata*

(*Cuspidariidae cont.*)
___ Ornate Cuspidaria, *Cardiomya ornatissima*

___ _____

Other Pelecypods

___ _____
___ _____
___ _____

Shells from Cephalopods

Paper Nautiluses (Argonautidae) ___
Note: Paper-like egg cases of octopus, *Argnauta*
___ Common Paper Nautilus, *Argonauta argo*
___ Brown Paper Nautilus, *Argonauta hians*

Spirulas (Spirulidae) ___
Note: Internal shells found on the beach
___ Common Spirula, *Spirula spirula*

Preparing Shells

Empty shells need little further attention. If you feel you must collect live specimens do so *sparingly*. The flesh is easily cleaned from most bivalves once the shells gape, as happens for example after a few hours in freshwater. For snails, remove the flesh by boiling in water for five minutes then, with a twisted or corkscrew motion, tease out

Shells

the inside. Resistant specimens may be tackled by hanging fleshy parts, letting the shell drop away when the attachment becomes fatigued.

Use bleach if it is necessary to clean the surface of the shell. (Avoid even the weakest acids, as these tend to dissolve the shell which is largely calcium carbonate.) To restore any shine or luster clouded by bleach, wipe the shell surface with a mixture of 1 part lighter fluid and 4 parts mineral oil.

Sometimes other special treatments are needed. See a good shell book for added suggestions.

Marine Invertebrates

Invertebrate listings, like those for bottom vegetation, are limited to distinct forms that may "catch the eye" of the observer. Except for a few Forams, whose skeletal remains may be seen on the blades of marine grasses and on hard surfaces, the vast array of one-celled animals, requiring a microscope to appreciate, is not considered. To a considerable extent the use of family groupings is abandoned and common names for general categories are used.

Forams ___
___ Turtle Grass and Button Forams, *Archais* spp.
___ Red Foram, *Hometrema rubrum*

Sponges ___
___ Cuban Reef Sponge, *Spongia obliqua*
___ Sheepswool Sponge, *Hippiospongia lachne*
___ Stinker Sponge, *Ircinia fasciculata*
___ Vase Sponge, *Ircinia campana*
___ Pillow Stinking Sponge, *Ircinia strobilina*
___ Candle Sponges, *Verongia* spp.
___ Finger Sponges, *Amphimedon* spp.
___ Don't-Touch-Me Sponge, *Neofibularia nolitangere*
___ Yellow Tube Sponge, *Aplisina fistularis*
___ Pipes-of-Pan Sponge, *Agelas schmidti*
___ Tube Sponges, *Callyspongia* spp.
___ Fire Sponge, *Tedania ignis*
___ Common Loggerhead Sponge, *Spheciospongia vesparum*
___ Giant Bowl Sponge, *Verangula gigantea*
___ Basket or Tub Sponge, *Xestospongia muta*
___ Boring Sponges, *Cliona* spp.
___ Golfball Sponges, *Tethya* spp.
___ Chicken Liver Sponge, *Chondrilla nucula*

___ _____
___ _____

Marine Invertebrates

Jellyfishes and Hydroids ___
___ Branched Hydroids, Several genera
___ Portuguese man-o-war, *Physalia physalis*
___ By-the-wind Sailor, *Vellela vellela*
___ Blue Button, *Porpita linneana*
___ Moon Jellyfish, *Aurelia aurita*
___ Cannonball Jellyfish, *Stomolophus meleagris*
___ Upsidedown Jellyfish, *Cassiopea xamachana or frondosa*
___ Sea Wasps, *Carybdea* spp.
___ _____

Comb or Balloon Jellies ___
___ Comb Jellies, *Mnemiopsis* spp.
___ _____

Flatworms ___
___ Crozier's Flatworm, *Pseudocerus crozieri*
___ _____

Ribbon Worms ___
___ Ribbon Worms, *Tubulanus* spp.
___ _____

Segmented or "True" Worms ___
___ Scale Worm, *Lepidonotus variabilis*
___ Fireworm, *Hermodice carunculata*
___ Bristle Worms, Several genera
___ Shaggy Parchment Tube Worm, *Onuphis magna*
___ Thread Worms, *Terebella* spp.

(Segmented or "True" Worms cont.)
___ Atlantic Palolo Worm, *Eunice schemacephala*
___ Common Fan Worm, *Sabellastarte magnifica*
___ Christmas Tree Worm, *Spirobranchus giganteus*
___ Miscellaneous Feather-duster and Fan worms, Sabellids and Serpulids, Several genera
___ _____
___ _____
___ _____

Sipunculids ___
___ Antillean Sipunculid, *Phascolosoma antillarum*
___ _____

Moss Animals ___
___ Common Bugula, *Bugula neritina*
___ Sargassum Sea Mat, *Membranipora tuberculata*
___ _____
___ _____

Sea Hares, Sea Slugs and Nudibranchs ___
___ Common Lettuce Slug, *Tridachia crispata*
___ Sea Hare, *Aplysia protea*
___ Other Sea Hares, Several genera
___ Rough Sea Slug, *Pleurobranchus aereolatus*
___ Nudibranchs, Several genera
___ _____
___ _____

Marine Invertebrates

Octopuses and Squids ___
- ___ Common Reef Octopus, *Octopus briareus*
- ___ Reef Squid, *Sepioteuthis sepioidea*
- ___ _____

Brine Shrimp ___
- ___ Brine Shrimp, *Artemia salina*

Barnacles ___
- ___ Goose Barnacles, *Lepas* spp.
- ___ Balanoid barnacles, *Balanus* spp.
- ___ Ribbed Barnacle, *Tetraclita stalactifera*
- ___ Fragile Barnacles, *Chthamalus* spp.
- ___ _____

Beach Fleas ___
- ___ Common Beach Hoppers, More than one genus
- ___ _____

Gribbles and Boat Roaches ___
- ___ Gribbles, More than one genus including wood borers
- ___ Boat Roaches, *Ligia* spp.
- ___ _____

Shrimps ___
- ___ Penaeid (Commercial) Shrimps, *Penaeus* spp.
- ___ Common Snapping Shrimp, *Synalpheus brevicarpus*
- ___ Other Snapping Shrimps, More than one genus

(Shrimps cont.)
- ___ Banded Coral Shrimp, *Stenopus hispidus*
- ___ Cleaning Shrimps, *Periclimenes* spp. and *Lysnata* spp.
- ___ Grass Shrimp, *Tozeuma carolinensis*
- ___ Other Grass Shrimps, *Crangon* spp.
- ___ _____
- ___ _____

Lobsters ___
- ___ Spiny Lobster, *Panulirus argus*
- ___ Spotted Spiny Lobster, *Panulirus guttatus*
- ___ Smooth-tailed Spiny Lobster, *Panulirus laevicauda*
- ___ Shovel-nose Lobster, *Scyllarides aequinoctialis*
- ___ _____

Callisanassids
- ___ Volcano Shrimp, *Callianassa* sp.

Porcelain Crabs ___
- ___ Poey's Porcellanid, *Megalobrachium poeyi*
- ___ Spotted Porcellain Crab, *Porcellana sayana*
- ___ _____

Hermit Crabs ___
- ___ Land Hermit Crab, *Coenobita clypeatus*
- ___ Giant Hermit Crab, *Petrochirus diogenes*

Marine Invertebrates

Hermit Crabs (cont.)
___ Pagurid Hermit Crabs, *Pagurus* spp.
___ _____ *Clibanarius* spp.
___ _____

Mole Crabs ___
___ Common Sand Fleas, *Emerita* spp.
___ Mole Crab, *Hippa cubensis*
___ Mole Crab, *Albrinea gibbesii*
___ _____

Sponge Crabs ___
___ Sponge Crabs, *Dromidia* spp. and *Domecia* spp.
___ _____

Purse Crabs ___
___ Purse Crab, *Persephone punctata*
___ _____

Box Crabs ___
___ Flamed Box Crab, *Callappa flammea*
___ Calico Crab, *Hepatus ephelticus*
___ _____

Swimming Crabs ___
___ Blue or Common Edible Crab, *Callinectes sapidus*
___ Ornate Crab, *Callinectes ornatus*
___ Spotted Swimming Crab, *Portunis sebae*
___ Sargassum Crab, *Portunus sayi*
___ _____

Xanthid Crabs ___
___ Coral Crab, *Carpilius corallinus*
___ Stone Crab, *Menippe mercenaria*
___ Calico Crab, *Eriphia gonagra*
___ _____

Grapsoid Crabs ___
___ Sally Lightfoot, *Grapsus grapsus*
___ Mangrove Tree Crab, *Goniopsis cruentata*
___ Mangrove Crab, *Aratus pisonii*
___ Wharf Crab, *Pachygraspus gracilis*
___ Gulf Weed Crab, *Planes minutus*
___ _____

Land Crabs ___
___ Great Land Crab, *Cardisoma guanhumi*
___ Black Land Crab, *Gecarcinus lateralis*
___ _____

Sand and Fiddler Crabs ___
___ Ghost Crab, *Ocypode quadrata*
___ Fiddler Crabs, *Uca* spp.
___ _____

Spider Crabs ___
___ Arrow Crab, *Stenorhynchus seticornus*
___ Spider Crab, *Libinia dubia*
___ Spiny Spider Crab or Caribbean King Crab, *Mithrax spinosissimus*
___ Green Reef Crab, *Mithrax sculptus*

(Spider Crabs cont.)
___ Decorator Crab, *Microphrys bicornutus*

_____

Elbow Crab ___
___ Elbow Crab, *Parthenope serrata*

_____

Mantis Shrimps ___
___ Common Rock Mantis Shrimp, *Gonodactylus oerstedi*
___ Other Mantis Shrimps, Several genera

_____

Horseshoe Crab ___
___ Horseshoe Crab, *Limulus polyphemus*

Starfishes ___
___ Two-spined Starfish, *Astropecten duplicatus*
___ Antilles Starfish, *Astropecten antillensis*
___ Nine-armed Luidia, *Luidia senegalensis*
___ Banded Luidia, *Luidia alternata*
___ Cushion Star, *Oreaster reticulatus*
___ Guilding's Star, *Ophidaster guildingii*
___ Common Comet Star, *Linckia guildingii*
___ Pentagon Star, *Asterina folium*
___ Thorny Starfish, *Echinaster sentus*

_____

Brittle Stars ___
___ Slimy Brittle Star, *Ophiomyxa flaccida*
___ Basket Star, *Astrophyton muricatum*
___ Mud Brittle Star, *Ophionephthys limicola*
___ Sponge Brittle Star, *Ophiothrix oerstedii*
___ Reticulate Brittle Star, *Ophionereis reticulata*
___ Ophiocomas, *Ophiocoma* spp.
___ Smooth Brittle Stars, *Ophioderma* spp.

_____

Sea Urchins and Sand Dollars ___
___ Slate-pencil Urchin, *Eucidaris tribuloides*
___ Long-spined Urchin, *Diadema antillarum*
___ Common Arbacia, *Arbacia punctulata*
___ Variegated Urchin, *Lytechinus variegatus*
___ Sea Egg, *Tripneustes esculentus*
___ Rock-boring Urchins, *Echinometra* spp.
___ Brown Sea Biscuit, *Clypeaster rosaceus*
___ Arrowhead Sand Dollar, *Encope michelini*
___ Key-hole Sand Dollar, *Mellita quinquiesperforata*
___ Six-hole Sand Dollar, *Mellita sexiesperforata*

222 Marine Invertebrates

(Sea Urchins and Sand Dollars cont.)
___ Reef Urchin, *Echinoneus cyclostomus*
___ Heart Urchin, *Moira atropus*
___ West Indian Sea Biscuit, *Meoma ventricosa*
___ _____
___ _____

Sea Lilies or Crinoids ___
___ Sea Lilies, Several genera
___ _____

Sea Cucumbers ___
___ Florida Sea Cucumber, *Holothuria floridana*
___ Donkey Dung Sea Cucumber, *Holothuria mexicana*
___ Golden Sea Cucumber, *Holothuria parvula*
___ Sticky-skinned Cucumber, *Euapta lappa*
___ Agassiz' Cucumber, *Actinopyga agassizii*
___ Four-sided Sea Cucumber, *Stichopus badionotus*
___ _____
___ _____

Sea Squirts ___
___ White Sponge Tunicate, *Didemnum candidum*
___ Mangrove Tunicate, *Ecteinascidia turbinata*
___ Black Tunicate, *Ascidia nigra*
___ Starred Amaroucium, *Amaroucium stellatum*
___ Divided Tunicates, *Styela partita*
___ Sandy-skinned Tunicate, *Molqula occidentalis*
___ Colorful Tunicate, *Clavelina oblonga*
___ _____
___ _____

Marine Bottom Plants

It takes an expert to "know" the bottom vegetation. Except for seagrasses, this vegetation, commonly filamentous in character, generally falls into three algal groups: Greens, Browns and Reds. The color is generally consistent within each group which helps the observer to sort the plants into broad categories.

The listing below is limited to a few forms sufficiently distinct to "catch-the-eye" of the general observer. The coralline algae, particularly the encrusting ones, are not so distinct but are included to encourage observers to be aware of their presence and their importance in reef building (see write-up of "Corals").

Directions for preparing mounts of algae are given at the end of this checklist.

Green Algae

Sea Lettuce, etc. (Ulvaceae) ___
___ Sea Lettuce, *Ulva lactuca*
___ Intestine Alga, *Enteromorpha flexuosa*

___ _____

_____(Anadyomenaceae) ___
___ _____,Anadyomene stellata

___ _____

_____ (Bryopsidaceae) ___
___ Bryopsis, *Bryopsis pennata*

___ _____

Cladophoras (Cladophoraceae) ___
___ Chaetomorphs, *Chaetomorpha* spp.
___ Cladophora, *Cladophora* spp.

___ _____

_____(Polyphysaceae) ___
___ Mermaid's Wine Glass, *Acetabularia crenulata*

___ _____

Bubble Algae (Valoniaceae) ___
___ Sea Bottle, *Ventricaria ventricosa*

223

(Valoniaceae cont.)
___ Green Bubble Alga,
   Dictyosphaeria cavernosa
___ _____
___ _____

_____ (Dasycladaceae) ___
___ _____, *Neomeris annulata*
___ _____

Caulerpas (Caulerpaceae) ___
___ Grape Alga, *Caulerpa racemosa*
___ _____, *Caulerpa sertularioides*
___ Other Caulerpas, *Caulerpa* spp.
___ _____
___ _____

Codiums (Codiaceae) ___
___ Codiums, *Codium* spp.
___ _____

Fans and Brushes (more than one family) ___
___ Mermaid's Fan, *Udotea flabellum*
___ Misc. fan-like algae, *Udotea* spp.
___ Soft-Fan Alga, *Avrainvillea nigricans*
___ Other Fan Algae, *Avrainvillea* spp.
___ Shaving Brush Alga, *Penicillus capitatus*
___ Other Brush Algae, *Penicillus* spp.
___ _____, *Rhipocephalus phoenix*
___ Cactus Halimeda, *Halimeda opuntia*

(Fans and Brushes cont.)
___ _____, *Halimeda incrassata*
___ Other Halimedas, *Halimeda* spp.
___ _____
___ _____

Green Algae in other families
___ _____
___ _____

Brown Algae

Dictyotas (Dictyotaceae) ___
___ Dictyotas, *Dictyota* spp.
___ _____, *Lobophora variegata*
___ Iridescent Banded Alga, *Stypopodium zonale*
___ Padina, *Padina sanctae-crusis*
___ _____

Sargassum (Sargassaceae) ___
___ Gulfweed (pelagic), *Sargassum fluitans* and *S. natans*
___ Gulfweed (attached), *Sargassum* spp.
___ _____

Turbinarians (Cystoseiraceae) ___
___ Turbinaria, *Turbinaria turbinata*
___ _____

Brown Algae in other families
___ _____
___ _____

Red Algae

_____ (Rhodomelaceae) ___
___ _____, *Laurencia* spp.
___ _____

Coralline Algae (Corallinaceae) ___
___ Branching Calcified Alga, *Amphiroa* ssp.
___ Red Coralline Alga, *Jania rubens*
___ Encrusting Coralline Algae, Several genera

___ _____
___ _____

Gracilarias (Gracilariaceae) ___
___ Gracilaria, *Gracilaria cervicornis*
___ Gracilaria, *Polycavernosa crassissima**
___ Gracilaria, *Polycavernosa debilis**

___ _____

_____ (Solieriaceae) ___
___ Eucheuma, *Eucheuma isiforme*

___ _____

Red Algae in other families

___ _____
___ _____

Seagrasses

Manatee and Shoal Grass (Potamogetonaceae) ___
___ Manatee Grass, *Syringodium filiforme*

___ _____

_____(Cymodoceaceae) ___
___ Shoal Grass, *Halodule beaudettei*

___ _____

Turtle and Related Grasses (Hydrocheritaceae) ___
___ Turtle Grass, *Thalassia testudinum*

___ _____

Mounting Algae

As you collect specimens, float them in seawater in a bucket or in plastic bags. After removing surface growths, debris, and excessive branches, mount the specimens while they are still fresh. For this you need sheets of heavy, good grade, high-rag-content paper that you can spread out on presswood and immerse in a shallow tray half-filled with seawater. Float each algal specimen onto the immersed paper, arranging the plant as desired. Tilt the board and paper slowly to allow

*These two species are also cultured using Gracilaria as the common name.

the water to run off gently, leaving the alga in place. The final arrangement of branches and filaments may be done with a weak stream of water from a pipette or medicine dropper.

The mounting paper, with algae in place, should now be placed either on thick blotting paper or on several layers of newspaper. After spreading unbleached muslin over the specimen, place more blotting material over it. Layers of such blotting material, specimens, and muslin can be stacked one above the other. To help drying, sheets of corrugated cardboard can be inserted in the stack at intervals. A board and a heavy weight should be put on top of the stack and the blotters or newspapers should be changed daily until the specimens are dry.

Soft, gelatinous, and finely-branched algae will usually adhere permanently to the mounting paper. Cartilaginous, wiry, or calcified specimens may require careful gluing. A little experience will soon yield well-mounted, decorative specimens. Since algal mounts can be useful for identifications and taxonomic work, the collection locale should be recorded.

Birds

(Many introduced exotics, park varieties, etcetera, are not included)

As is so often experienced, differences in life stages (as for gulls and terns) and groups of similar, closely related species (amongst the warblers, for example) complicate the observer's efforts. You may spot twenty or thirty of these birds rather quickly but it will take an added effort to identify twice as many unless you are an ardent birder. Since many birds are very mobile you may see some in *unexpected* localities. Such new sitings are part of the excitement of birding. Also, don't be surprised if you encounter established populations of escaped pets, particularly certain parrots and parakeets that have found the local surroundings *to their liking*. Some that have escaped and are well established are included in the list.

Though much is made of the obvious ability of birds to extend their range, it is noteworthy that isolation and the consequent evolution into separate species has occurred in a few instances. Interesting examples are the separate oriole species of Montserrat, St. Lucia and Martinique and the native parrot species unique to Dominica, St. Lucia and St. Vincent. You can prove you are an avid birder by searching for the distinct parrots in the mountain forests.

Except for local names there is sufficient professional agreement as to the identity of birds, and even of the families involved, to allow listing them by common names only.

M before a listing indicates a species that is most likely to be seen during its migration; **W** indicates the species is primarily a winter resident or both a migrant and winter visitor; the remainder are resident to a greater or lesser extent.

228 *Birds*

Grebes ___
   ___ Pied-billed Grebe
___ _____

Shearwaters and Petrels ___
   ___ Audubon's Shearwater
___ _____

Tropicbirds ___
   ___ Red-billed Tropicbird
   ___ White-tailed Tropicbird

Boobies ___
   ___ Brown Booby
   ___ Red-footed Booby
___ _____

Pelicans ___
   ___ Brown Pelican
___ _____

Cormorants ___
   ___ Double-crested Cormorant
___ _____

Frigatebird ___
   ___ Magnificent Frigatebird

Herons and Bitterns ___
   ___ Great Blue Heron
   ___ Great White Heron (color phase of Great Blue)
   ___ Green-backed Heron
   ___ Little Blue Heron
   ___ Cattle Egret
   ___ Great Egret
   ___ Snowy Egret

(Herons and Bitterns cont.)
   ___ Tricolored Heron
   ___ Black-crowned Night Heron
   ___ Yellow-crowned Night Heron
   ___ Least Bittern
___ _____

Ibises ___
   ___ Glossy Ibis
___ _____

Flamingos ___
   ___ Greater Flamingo

Ducks ___
   ___ West Indian Whistling Duck
   ___ White-cheeked Pintail
   ___ Blue-winged Teal
   ___ American Widgeon
   ___ Northern Shoveler
   ___ Wood Duck
W ___ Lesser Scaup
W ___ Ring-necked Duck
   ___ Ruddy Duck
___ _____
___ _____

Vultures ___
   ___ Turkey Vulture

Hawks ___
   ___ Hook-billed Kite
   ___ Red-tailed Hawk
   ___ Broad-winged Hawk
W ___ Northern Harrier
___ _____

Birds 229

Osprey ___
W___ Osprey

Falcons ___
W___ Peregrine Falcon
W___ Merlin
___ American Kestrel

Junglefowl, Quail and
    Guineafowl ___
___ Helmeted Guineafowl
___ Northern Bobwhite

Rails and Coots ___
___ Clapper Rail
W___ Sora
___ Common Moorhen
___ American Coot
___ Caribbean Coot

Oystercatchers ___
___ American Oystercatcher

Plovers ___
M___ Semipalmated Plover
___ Snowy Plover
___ Wilson's Plover
___ Killdeer
M___ Lesser Golden Plover
W___ Black-bellied Plover

Stilts ___
___ Black-necked Stilt

(Stilts cont.)
___ Common Stilt

Woodcock and Sandpipers ___
W___ Ruddy Turnstone
W___ Common Snipe
W___ Whimbrel
W___ Spotted Sandpiper
M___ Solitary Sandpiper
W___ Greater Yellowlegs
W___ Lesser Yellowlegs
___ Willet
M___ Pectoral Sandpiper
___ Peep Sandpipers*
M   White-rumped
W   Least
M   Semi-palmated
W   Western
W___ Sanderling
W___ Short-billed Dowitcher
M___ Stilt Sandpiper

Gulls ___
W___ Ring-billed Gull
___ Laughing Gull

Terns ___
___ Common Tern
___ Roseate Tern
___ Bridled Tern
___ Sooty Tern
___ Least Tern

*Not distinguishable in the field.

230 Birds

(Terns cont.)
___ Royal Tern
___ Sandwich Tern
M___ Black Tern
___ Brown Noddy

Pigeons and Doves ___
___ White-crowned Pigeon
___ Scaly-naped Pigeon
___ Rock Dove
___ Violet-eared Dove
___ Zenaida Dove
___ Common Ground Dove
___ Ruddy Quail-dove
___ Bridled Quail-dove
___ Grenada Dove

Parrots and Parakeets ___
___ Budgerigar
___ Canary-winged Parakeet
___ Imperial Parrot (Dominica only)
___ St. Vincent Parrot (St. Vincent only)
___ St. Lucia Parrot (St. Lucia only)
___ Monk Parakeet

Cuckoos ___
___ Mangrove Cuckoo
___ Yellow-billed Cuckoo
___ Puerto Rican Lizard Cuckoo
___ Smooth-billed Ani

Owls ___
___ Common Barn Owl
___ Puerto Rican Screech Owl
___ Short-eared Owl
___ Burrowing Owl

Nightjars ___
___ White-tailed Nightjar
___ Common Nighthawk

Swifts ___
___ Lesser Antillean Swift
___ Black Swift

Hummingbirds ___
___ Puerto Rican Emerald Hummingbird
___ Antillean Mango
___ Green Mango
___ Purple-throated Carib
___ Green-throated Carib
___ Antillean Crested Hummingbird
___ Blue-headed Hummingbird
___ Rufus-breasted Hermit

Kingfishers ___
___ Ringed Kingfisher
W___ Belted Kingfisher

Todies ___
___ Puerto Rican Tody

Woodpeckers ___
   ___ Puerto Rican Woodpecker
   ___ Guadeloupe Woodpecker
___ _____

Flycatchers ___
   ___ Tropical Kingbird
   ___ Gray Kingbird
   W___ Fork-tailed Flycatcher
   ___ Loggerhead Kingbird
   ___ Puerto Rican Flycatcher
   ___ Rusty-tailed Flycatcher
   ___ Grenada Flycatcher
   ___ Lesser Antillean Flycatcher
   ___ Stolid Flycatcher
   ___ Lesser Antillean Pewee
   ___ Yellow-bellied Elaenia
   ___ Caribbean Elaenia
___ _____

Swallows ___
   ___ Caribbean Martin
   M___ Bank Swallow
   W___ Barn Swallow
   ___ Cave Swallow
___ _____

Wrens ___
   ___ House Wren
___ _____

Mocking Birds and Thrashers ___
   ___ Northern Mockingbird
   ___ Tropical Mockingbird
   ___ Scaly-breasted Thrasher
   ___ Pearly-eyed Thrasher
   ___ Trembler

(Mocking Birds and Thrashers cont.)
   ___ White-breasted Thrasher
___ _____

Thrushes ___
   ___ Red-legged Thrush
   ___ Cocoa Thrush
   ___ Forest Thrush
   ___ Bare-eyed Thrush
   ___ Rufous-throated Solitaire
___ _____

Starlings ___
   ___ European Starling

Vireos ___
   ___ Puerto Rican Vireo
   ___ Black-whiskered Vireo
___ _____

Warblers ___
   W___ Black and White Warbler
   W___ Prothonotary Warbler
   W___ Northern Parula
   ___ Yellow Warbler
   W___ Cape May Warbler
   W___ Black-throated Blue Warbler
   W___ Yellow-rumped (Myrtle) Warbler
   ___ Adelaide's Warbler
   M___ Blackpoll Warbler
   W___ Prairie Warbler
   ___ Plumbeous Warbler
   W___ Northern Waterthrush
   W___ Louisiana Waterthrush
   W___ Ovenbird
   W___ Common Yellowthroat

232   Birds

(Warblers cont.)
W___ American Redstart
___ _____
___ _____

Honeycreepers ___
___ Bananaquit
___ _____

Tanagers ___
___ Antillean Euphonia
___ Stripe-headed Tanager
___ Puerto Rican Tanager
___ Hooded Tanager
___ _____

Cardinals and Allies ___
___ Streaked Saltator
W___ Indigo Bunting

Blackbirds and Orioles ___
___ Shiny Cowbird
___ Greater Antillean Grackle
___ Carib Grackle
___ Black-cowled Oriole
___ Montserrat Oriole (Montserrat only)
___ St. Lucia Oriole (St. Lucia only)
___ Martinique Oriole (Martinique only)
___ Troupial
___ Yellow-shouldered Blackbird
___ _____

Waxbills ___
___ Orange-cheeked Waxbill
___ Black-rumped Waxbill
___ Bronze Mannikin
___ Chestnut Mannikin
___ Red Avadavat
___ _____

Finches and Sparrows ___
___ Yellow Grass Finch
___ Yellow-bellied Seedeater
___ Puerto Rican Bullfinch
___ Lesser Antillean Bullfinch
___ Saffron Finch
___ Yellow-faced Grassquit
___ Black-faced Grassquit
___ Grasshopper Sparrow
___ _____
___ _____

Flowers,* Shrubs, and Vines

In the remaining listings for flowers and for trees, the coverage again is quite limited.† The emphasis is on the showy and otherwise interesting forms. Does this involve serious oversights?—Yes! But, due to the overwhelming diversity of the flora, including many forms that are ecologically very important yet relatively obscure, only professional botanists, armed with specialized references, can hope to do more. In general the agricultural crops are not listed unless they are unique to the tropics.

In addition to selecting what to point out, arbitrary to say the least, there are a number of complicating factors:

- variations in response to the environment can be confusing
- many species have several color forms and other variants
- hybrids occur and variants have been selectively cultivated, to the delight of flower lovers and the advance of agriculture
- there is little consistency as to the common names used, not to mention the difficulties with synonyms for the scientific names.

In coping with common names I tend to avoid the more local ones, favoring those which, in my opinion, may be familiar in a broader sense. Where available I also favor names that are a bit descriptive. Some of the common names you will encounter encompass more than

*Nature observers tend to use the word *flower* for the overall colorful inflorescence of a plant, though strictly speaking this may include modified leaves or other non-floral parts as in the colorful terminal leaves of Poinsettias and Bougainvilleas.

†Fungi, Lichens and Mosses are not dealt with in this listing.

234 Flowers, Shrubs, and Vines

one species and sometimes even more than one genus. For example "Lady of the Night" may refer to any plant with nocturnally fragrant flowers.

Letters preceding each entry indicate whether the species is **N** = native, **X** = exotic, in other words, introduced, **XN** = exotic that has become naturalized, **N & X** is used where there are several species some of which are native and some exotic. Perhaps more of the introductions have become naturalized than indicated in the lists.

There is an element of uncertainty in such designations. While historical records and other tangible evidence may indicate that various species have been introduced, there are some for which the accounts are not altogether reliable. And there are others for which the possible time of introduction is so far in the past that we must wonder whether they are native or were brought in by early inhabitants of the region. That authorities may disagree hinges not only on these uncertainties but also on the fact that a species native to a given part of the region may have been introduced into another.

Ground and Climbing Ferns (Several families; many genera and species) ___
N   ___ Mangrove Fern, *Acrostichum aureum*
N   ___ Print Fern, *Pityrogramma calomelanos*
N   ___ Maidenhair Ferns, *Adiantum* spp.
N   ___ Polypody Ferns, *Polypodium* spp.
N&X   ___ Sword Ferns, *Nephrolepis* spp.
___ _____

Grass Family (Poaceae or Gramineae) ___
X   ___ Sugar Cane, *Saccharum officinarum*
X   ___ Corn, *Zea mays*
N   ___ Sandspur, *Cenchrus echinatus*
X   ___ Johnson Grass, *Sorghum halepense*
X   ___ Rice, *Oryza sativa*
XN   ___ Job's Tears, *Coix lacryma-jobi*
X   ___ Guinea Grass, *Panicum maximum*
___ _____

Flowers, Shrubs, and Vines

Sedges (Cyperaceae) ___
X ___ Umbrella Sedge, *Cyperus alternifolius*
___ _____
___ _____

Arums (Araceae) ___
X ___ Anthurium, *Anthurium andraeanum*
N&X ___ Other Anthuriums, *Anthurium* spp.
X ___ Common Caladium, *Caladium bicolor*
X ___ Other Caladiums, *Caladium* spp. (including many hybrids)
X ___ Spathiphyllum, *Spathiphyllum* spp.
N ___ Elephant Ear, *Philodendron giganteum*
N ___ Other Philodendrons, *Philodendron* spp.
X ___ Swiss Cheese Plant, *Monstera deliciosa*
X ___ Tannia, *Xanthosoma violacea*
X ___ Wild Taro, *Xanthosoma* spp.
X ___ Dasheen or Taro, *Colocasia esculenta*
X ___ Golden Pothos, *Epipremnum aureum*
X ___ Dumb Cane, *Dieffenbachia maculata*
___ _____
___ _____

Bromeliads (Bromeliaceae) ___
X ___ Pineapple, *Ananas comosus*
N ___ Bromelia, *Bromelia pinquin*
N ___ Other Bromeliads (Several genera)
N ___ Old Man's Beard, *Tillandsia recurvata*
N ___ Other Tillandsias, *Tillandsia* spp.
___ _____
___ _____

Spiderwort Family (Commelinaceae) ___
X ___ Wandering Jew, *Zebrina pendula*
N ___ Oyster Plant, *Rhoeo spathacea*
X ___ Purple Queen, *Setcreasea pallida*
X ___ Callisia, *Callisia fragrans*
___ _____
___ _____

Pickerel Weed Family (Pontederiaceae) ___
X ___ Water Hyacinth, *Eichornia crassipes*
___ _____

Lilies (Liliaceae) ___
X ___ Glory Lily, *Gloriosa superba*
X ___ Aloe, *Aloe barbadense*
___ _____

Flowers, Shrubs, and Vines

Amaryllis Family
   (Amaryllidaceae) ___
N   ___ Amaryllis or Easter Lily, *Hippeastrum puniceum*
X   ___ Amazon Lily, *Eucharis grandiflora*
N   ___ Spider Lily, *Hymenocallis caribaea*
X   ___ Crinum Lilies, *Crinum* spp.

___ _____

Agave Family (Agavaceae) ___
X   ___ Sand Dragon, *Dracaena fragrans*
X   ___ Ti Plant, *Cordyline terminalis*
X   ___ Guana Tail, *Sansevieria guineensis*
X   ___ Spanish Bayonets, *Yucca* spp.
X&N ___ Century Plants, *Agave* spp.

___ _____
___ _____

Yams (Dioscoreacea) ___
X&N ___ Yams, *Dioscorea* spp.

___ _____

Irises (Iridaceae) ___
X   ___ Walking Iris, *Neomarica* spp.

___ _____

Strelitzias (Strelitziaceae) ___
X   ___ Bird of Paradise, *Strelitzia reginae*

___ _____

Heliconias (Heliconiaceae) ___
N   ___ Common Heliconia, *Heliconia caribaea*
X   ___ Hanging Lobster Claw, *Heliconia rostrata*
XN ___ Other Heliconias, *Heliconia* spp.

___ _____
___ _____

Gingers (Zingiberaceae) ___
X   ___ Common Ginger, *Zingiber officinale*
X   ___ Red Ginger, *Alpinia purpurata*
X   ___ Shell Ginger, *Alpinia zerumbet*
X   ___ Torch Ginger, *Nicolaia elatior*
X   ___ Spiral Ginger, *Costus speciosus*
X   ___ White Ginger, *Hedychium coronarium*

___ _____
___ _____

Cannas (Cannaceae) ___
X   ___ Canna Lily or Indian Shot, *Canna indica*
X   ___ Other Cannas, spp. and hybrids

___ _____
___ _____

Arrowroots (Marantaceae) ___
X   ___ Arrowroot, *Maranta arundinacea*
X   ___ Calatheas, *Calathea* spp.

___ _____

Flowers, Shrubs, and Vines

Orchids (Orchidaceae) ___ (Many genera and species plus varieties; hundreds may be found in the islands)
N ___ Christmas Orchid, *Epidendrum ciliare*
N ___ Cockleshell Orchid, *Epidendrum cochleatum*
X ___ Vanilla Bean, *Vanilla planifolia*
N ___ Dancing Ladies, *Oncidium* spp.
___ _____
___ _____
___ _____

Smartweeds (Polygonaceae) ___
X ___ Coral Vine, *Antigonon leptopus*
___ _____
___ _____

Salicornias (Chenopodiaceae) ___
N ___ Glasswort, *Salicornia bigelowii*
___ _____

Amaranths (Amaranthaceae) ___
X ___ Beafsteak Plant, *Irisine herstii*
X ___ Devil's Horse Whip, *Achyranthus indica*
X ___ Calabash Rose or Marguerite, *Gomphrena globosa*
___ _____

Four-O-Clocks (Nyctaginaceae) ___
X ___ Bougainvillea, *Bougainvillea spectabilis*
X ___ Other Bougainvilleas, *Bougainvillea* spp.
X ___ Four O'Clock, *Mirabilis jalapa*
X ___ Hogweed, *Boerhavia coccinea*
___ _____

Piper Family (Piperaceae) ___
N ___ Black Wattle, *Piper amalgo*
___ _____

Saltworts (Bataceae) ___
N ___ Saltwort, *Batis maritima*
___ _____

Pokeweed Family (Phytolaccaceae) ___
N ___ Congo Root, *Petiveria alliacea*
N ___ Hoop Vine, *Trichostigma octandrum*
___ _____

Purslanes (Aizoaceae) ___
N ___ Seaside Purslane, *Sesuvium portulacastrum*
___ _____

Portulacas (Portulacaceae) ___
XN ___ Pussley, *Portulaca oleracea*
___ _____

Flowers, Shrubs, and Vines

Water Lilies (Nymphaeaceae) ___
X ___ Sacred Lotus, *Nelumbo nucifera*
X ___ Water Lilies, *Nymphaea* spp.

___ _____

Caper Family (Capparaceae) ___
X ___ Spider Plant, *Cleome speciosa*
X ? ___ Sticky Cleome, *Cleome viscosa*

___ _____

Stonecrop Family (Crassulaceae) ___
XN ___ Bryophyllum or Leaf of Life, *Kalanchoe pinnata*

___ _____

Chrysobalanus Family (Chrysobalanaceae) ___
N ___ Coco Plum, *Chrysobalanus icaco*

___ _____

Pea Family (Fabaceae) ___
X ___ Peanut, *Arachis hypogaea*
N ___ Nicker Bean, *Caesalpinia bonduc*
X ___ Dwarf Poinciana, *Caesalpinia pulcherrima*
N ___ Rabbit Vine, *Teramnus labialis*
N ___ Rabbit Vine, *Centrosema virginianum*
N ___ Yellow Rattle, *Crotalaria retusa*
N ___ Crab's Eye, *Abrus precatorius*

(Fabaceae cont.)
X ___ Jade Vine, *Strongylodon macrobotrys*
X ___ Pigeon Pea, *Cajanus cajan*
X ___ Sensitive Plant, *Mimosa pudica*
N ___ Seaside Bean, *Canavalia maritima*
X ___ Butterfly Pea, *Clitoria ternatea*
N ___ Necklace Pod, *Sophora tomentosa*
N ___ Stinking Weed, *Cassia occidentalis*
X ___ Powder Puff, *Calliandra inaequilatera*
X ___ Sea Heart, *Entada gigas*

___ _____
___ _____

Quassia Family (Simaroubaceae)
N ___ Bay Cedar, *Suriana maritima*
X ___ Bitterwood, *Quassia amara*

___ _____

Malpighia Family (Malpighiaceae) ___
N ___ West Indian Holly, *Malpighia coccigera*
N ___ Golden Vine, *Stigmaphyllon* spp.
X ___ Galphimia, *Galphimia glauca*

___ _____
___ _____

Flowers, Shrubs, and Vines

Spurges (Euphorbiaceae) ___
- X ___ Castor Bean, *Ricinus communis*
- X ___ Chenille Plant, *Acalypha hispida*
- X ___ Garden Croton, *Codiaeum variegatum*
- N ___ Maran, *Croton astroites*
- N ___ Other Crotons, *Croton* spp.
- XN ___ Cassava, *Manihot esculenta*
- N ___ Christmas Candle, *Pedilanthus tithymaloides*
- N ___ Physic Nut, *Jatropha curacas*
- X ___ Poinsettia, *Euphorbia pulcherrima*
- X ___ Crown of Thorns, *Euphorbia milii*
- X ___ Castor Bean, *Ricinus communis*
- ___ ___
- ___ ___

Mallows (Malvaceae) ___
- X ___ Hibiscus, *Hibiscus rosa-sinensis*
- X ___ Coral Hibiscus, *Hibiscus schizopetalus*
- X ___ Okra, *Hibiscus esculentus*
- X ___ Sorrell, *Hibiscus sabdariffa*
- X ___ Other Hibiscuses, *Hibiscus* spp. (including many hybrids)
- X ___ Turk's Cap or Sleeping Hibiscus, *Malvaviscus arboreus*

(Malvaceae cont.)
- X ___ Sea Island Cotton, *Gossypium barbadense*
- N ___ Broom, *Sida acuta*
- ___ ___
- ___ ___

Turnave Family (Turneraceae) ___
- N ___ Sage Rose, *Turnera ulmifolia*
- ___ ___

Passion Flowers (Passifloraceae) ___
- X ___ Red Passion Flower, *Passiflora rubra*
- X ___ Passion Fruit, *Passiflora edulis*
- N ___ Bell Apple, *Passiflora laurifolia*
- XN ___ Other Passion Flowers, *Passiflora* spp. (numerous species may be found)
- ___ ___
- ___ ___

Begonias (Begoniaceae) ___
- X ___ Angel Wing Begonia, *Begonia corallina*
- N ___ Scented Begonia, *Begonia minor*
- ___ ___

Cacti (Cactaceae) ___
- N ___ Prickly Pear, *Opuntia dillenii*
- N ___ Sucker cactus, *Opuntia repens*
- N ___ Other Opuntias, *Opuntia* spp.

Flowers, Shrubs, and Vines

(Cactaceae cont.)
- N ___ Dildo, *Cephalocereus royenii*
- N ___ Turk's Head Cactus, *Melocactus intortus*
- N ___ Night Blooming Cereus, *Hylocereus undatus*
- N ___ Currant Cactus, *Rhipsalis baccifera*
- X ___ Rose cacti, *Pereskia* spp.
- ___ _____
- ___ _____

Loosestrife Family (Lythraceae) ___
- X ___ Cupheas, *Cuphea* spp.
- X ___ Crepe Myrtle, *Lagerstroemea indica*
- ___ _____

Parsley Family (Apiaceae) ___
- X ___ Anise, *Pimpinella anisum*
- ___ _____

Plumbagos (Plumbaginaceae) ___
- X ___ Plumbago, *Plumbago auriculata*
- N ___ Wild Plumbago, *Plumbago scandens*
- ___ _____

Olives (Oleaceae) ___
- XN ___ Azores jasmine, *Jasminum fluminense*
- XN ___ Hairy jasmine, *Jasminum multiflorum*
- X ___ Royal Jasmine, *Jasminum grandiflorum*
- X ___ Other Jasmines, *Jasminum* spp.
- ___ _____

Dogbanes (Apocynaceae) ___
- X ___ Allamanda, *Allamanda cathartica*
- X ___ Beaumontia or Herald's Trumpet, *Beaumontia grandiflora*
- X ___ Oleander, *Nerium oleander*
- XN ___ Periwinkle (Old Maid), *Catharanthus roseus*
- N ___ Wild Allamanda, *Urechites lutea*
- XN ___ False Gardenia, *Tabernaemontana coronaria*
- X ___ Natal Plum, *Carissa macrocarpa*
- ___ _____

Milkweeds (Asclepiadaceae) ___
- N ___ Red Head, *Asclepias curassavica*
- X ___ Madagascar Jasmine, *Stephanotis floribunda*
- XN ___ Giant Milkweed, *Calotropis procera*
- XN ___ Rubber Vine, *Cryptostegia grandiflora*
- ___ _____
- ___ _____

Morning Glories (Convolvulaceae) ___
- X ___ Morning Glory, *Ipomoea purpurea*
- N ___ Bush Morning Glory or Potato Bush, *Ipomea fistulosa*
- N ___ Beach Morning Glory, *Ipomoea pes-caprae*

Flowers, Shrubs, and Vines

(Convolvulaceae cont.)
- X ___ Sweet Potato, *Ipomoea batatus*
- X ___ Wood Rose, *Merremia tuberosa*
- N ___ Love Vine or Dodder, *Cuscuta americana*
- N ___ Bindweed, *Jacquemontia pentantha*
- N ___ Speedwell, *Evolvulus alsinoides*
- ___ _____

Verbenas (Verbenaceae) ___
- N ___ Yellow Sage, *Lantana camara*
- N ___ Other Lantanas, *Lantana* spp.
- X ___ Bleeding Heart, *Clerodendrum thomsoniae*
- X ___ Pagoda Flowers, *Clerodendrum* spp.
- XN ___ Petrea, *Petrea volubilis*
- X ___ Congea, *Congea tomentosa*
- N ___ Vervain, *Stachytarpheta jamaicensis*
- N ___ Skyflower, *Duranta repens*
- X ___ Chinaman's Hat, *Holmskioldia sanguinea*
- ___ _____
- ___ _____

Mints (Lamiaceae) ___
- X ___ Lion's Ear, *Leonotis nepetifolia*
- N ___ Balsam, *Ocium micranthum*

(Lamiaceae cont.)
- XN ___ Wild Thyme, *Coleus amboinicus*
- X ___ Miscellaneous Coleus, *Plectranthus blumei* & hybrids
- ___ _____
- ___ _____

Nightshades (Solanaceae) ___
- N ___ Cakalaka Berry, *Solanum polyganum*
- X ___ Aubergine or Eggplant, *Solanum melongena*
- XN ___ Wiri Wiri, *Capsicum frutescens*
- XN ___ Peppers, *Capsicum* spp.
- X ___ Cup of Gold, *Solandra nitida*
- N ___ Lady of the Night, *Brunsfelsia americana*
- X ___ Thorn Apple, *Datura stramonium*
- X ___ Angel's Trumpet, *Brugmansia X Candida*
- ___ _____
- ___ _____

Figworts (Scrophulariaceae) ___
- X ___ Firecracker, *Russelia equisetiformis*
- N ___ Angelonias, *Angelonia* spp.
- ___ _____

Bignonias (Bignoniaceae) ___
- X ___ Flame Vine, *Pyrostegia venusta*

242 Flowers, Shrubs, and Vines

(Bignoniaceae cont.)
XN ___ Cat's Claw, *Macfadyena unguis-cati*
N ___ Garlic Vine, *Cydista aequinoctialis*
X ___ Cape Honeysuckle, *Tecomaria capensis*

___ _____

Bladderworts (Lentibulariaceae) ___
N ___ Bladderworts, *Utricularia* spp.

___ _____

Acanthus Family (Acanthaceae) ___
X ___ Clock Vine, *Thunbergia grandiflora*
X ___ White Nightshade, *Thunbergia fragrans*
X ___ Black-eyed Susan Vine, *Thunbergia alata*
X ___ King's Mantle, *Thunbergia erecta*
X ___ White Shrimp Plant, *Justica brandegeana*
X ___ Other Shrimp Plants, *Justica* spp.
X ___ Garden Coral, *Odontonema strictum*
X ___ Blue Sage, *Eranthemum nervosum*
X ___ Golden Eranthemum, *Pseuderanthemum reticulatum*

___ _____
___ _____

Madders (Rubiaceae) ___
X ___ Jungle flame, *Ixora coccinea*
X ___ Ixoras, *Ixora* spp.
N ___ White Broom, *Borreria verticillata*
N ___ Scarlet Bush, *Hamelia patens*
X ___ Mussaenda, *Mussaenda erythrophylla*
N ___ Bell Flower, *Portlandia grandiflora*
N ___ Snowberry, *Chiococca alba*
N ___ Wild Coffee, *Psychotria nervosa*

___ _____

Gourds (Cucurbitaceae) ___
XN ___ Maiden Apple or Carilla, *Momordica charantia*
X ___ Watermelon, *Citrullus lanatus*
X ___ Pumpkin, *Cucurbita maxima*
XN ___ Christophene, *Sechium edule*

___ _____

Goodenia Family (Goodenaceae) ___
N ___ Ink Berry, *Scaevola plumieri*

___ _____

Composites (Asteraceae or Compositae) ___
X ___ Orange Glow Vine, *Senecio confusus*

Flowers, Shrubs, and Vines

X   ___ Consumption Weed, *Emilia coccinea*
N   ___ Whitehead, *Parthenium hysterophorus*
N   ___ Inflamation Bush, *Vernonia cinerea*
N   ___ Spanish Needle, *Bidens pilosa*

___ _____
___ _____

Trees

While all trees produce flowers, it is only in the tropics that in all seasons there is an abundance of trees with showy blossoms. These are commonly referred to as "flowering trees." Inevitably these "flowering" types predominate in a listing of highlights as presented herewith.

The general comments in the introduction to "Flowers, Shrubs and Vines" apply under this heading.

Ferns and Fern Allies (Several families, genera and species) ___
N ___ Tree Ferns, *Cyathea* spp.
___ _____
___ _____

Cycads (Cycadaceae) ___
X ___ Queen Sago, *Cycas circinalis*
X ___ Japanese cycad, *Cycas revoluta*
X&N ___ Zamias, *Zamia* spp.
___ _____

Pines (Pinaceae) ___
N ___ Caribbean Pine, *Pinus caribaea*
___ _____

Yews (Podocarpaceae) ___
N ___ Podocarp, *Podocarpus coriaceus*
___ _____

Auracaria Family (Araucariaceae) ___
X ___ Norfolk Island Pine, *Araucaria heterophylla*
___ _____

Grass Family (Poaceae or Gramineae) ___
X ___ Bamboo, *Bambusa vulgaris*
___ _____

Palms (Arecaceae or Palmae) ___
XN ___ Coconut Palm, *Cocos nucifera*
X ___ Queen Palm, *Syagrus romanzoffiana*

Trees

(Aracaceae or Palmae cont.)
- X ___ Date Palm, *Phoenix dactylifera*
- X ___ Areca Palm, *Chrysalidocarpus lutescens*
- X ___ Christmas Palm, *Veitchia merrillii*
- X ___ Fiji Fan Palm, *Pritchardia pacifica*
- X ___ Chinese Fan Palm, *Livistona chinensis*
- N ___ Royal Palm, *Roystonea regia*
- N ___ Cabbage Palm, *Roystonea oleracea*
- N ___ Sabal Palms, *Sabal* spp.
- N ___ Broom Palms, *Coccothrinax* spp.
- ___ _____
- ___ _____

Pandanus Family (Pandanaceae) ___
- X ___ Common Screwpine, *Pandanus utilis*
- X ___ Other Screwpines, *Pandanus* spp.
- ___ _____

Bananas (Musaceae) ___
- X ___ Bananas and Plantains, *Musa* spp.
- ___ _____
- ___ _____

Strelitzias (Strelitziaceae) ___
- X ___ White Bird of Paradise, *Strelitzea nicolai*

(Strelitziaceae cont.)
- X ___ Traveler's Tree, *Ravenala madagascariensis*
- ___ _____

Casuarinas (Casuarinaceae) ___
- X ___ Australian Pine or Casuarina, *Casuarina equisetifolia*

Mulberry Family (Moraceae) ___
- X ___ Breadfruit and Breadnut Trees, *Artocarpus communis*
- X ___ Jackfruit Tree, *Artocarpus heterophyllus*
- N ___ Trumpet Tree, *Cecropia peltata*
- X ___ Rubber Plant, *Ficus elastica*
- X ___ Banyan Tree, *Ficus benghalensis*
- N ___ Short-leaf Fig, *Ficus citrifolia*
- N&X ___ Other Ficus, *Ficus* spp.
- ___ _____

Buckwheat Family (Polygonaceae) ___
- N ___ Sea Grape, *Coccoloba uvifera*
- XN ___ Long John Tree, *Triplaris americana*
- N ___ Redwood, *Coccoloba diversifolia*
- ___ _____

Trees

Four-O'clock Family (Nyctaginaceae)
N ___ Loblolly, *Pisonia subcordata*
N ___ Black Mampoo, *Guapira fragrans*

———————————————

Custard Apples (Annonaceae) ___
N ___ Custard Apple, *Annona reticulata*
N ___ Soursop, *Annona muricata*
N ___ Sugar Apple or Sweetsop, *Annona squamosa*
N ___ Blackbarb, *Guatteria caribaea*
X ___ Ylang Ylang Tree, *Cananga odorata*

———————————————

Nutmegs (Myristicaceae) ___
X ___ Nutmeg Tree, *Myristica fragrans*

———————————————

Laurels (Lauraceae) ___
X ___ Avocado, *Persea americana*

———————————————

Canella Family (Canellaceae) ___
N ___ White Cinnamon, *Canella winterana*

———————————————

Mangosteen Family (Clusiaceae or Guttiferae) ___
N ___ Autograph Tree, *Clusia rosea*

(Clusiaceae or Guttiferae cont.)
N ___ Mammee Apple, *Mammea americana*
N ___ Santa Maria, *Calophyllum brasiliensis*

———————————————

Capers (Capparaceae) ___
N ___ Jamaica Caper, *Capparis cynophallophora*
N ___ Limber Caper, *Capparis flexuosa*
N ___ Other Capers, *Capparis* spp.

———————————————

Horseradish Tree Family (Moringaceae) ___
X ___ Horseradish Tree, *Moringa pterygosperma*

———————————————

Pea Family (Fabaceae) ___
X ___ Sweet Acacia, *Acacia farnesiana*
N ___ Casha, *Acacia macracantha*
N&XN ___ Other Cashas, *Acacia* spp.
X ___ Golden Shower Tree, *Cassia fistula*
X ___ Pink Cassia, *Cassia javanica*
X ___ Christmas Candle, *Cassia alata*
N,X&XN ___ Other Cassias, *Cassia* spp.
N ___ Sweet Pea, *Inga laurina*
X ___ Flamboyant Tree or Royal Poinciania, *Delonix regia*

(Fabaceae cont.)

X	___ Mesquite, *Prosopis juliflora*		
X	___ Pride of India, *Peltophorum pterocarpum*		
XN	___ Rain Tree, *Samanea saman*		
X	___ Logwood, *Haematoxylum campechianum*		
X	___ Orchid Trees, *Bauhinia* spp.		
XN	___ Tamarind, *Tamarindus indica*		
N	___ Pig Turd, *Andira inermis*		
XN	___ Wild Tamarind, *Leucaena leucocephala*		
X	___ Tiger's Claw, *Erythrina variegata*		
X	___ Coral Tree, *Erythrina crista-galli*		
X	___ Earpod Tree, *Enterolobium cyclocarpum*		
X	___ Parkinsonia, *Parkinsonia aculeata*		
X	___ Bead Tree, *Adenanthera pavonina*		
X	___ Mother of Cocoa, *Gliricidia sepium*		
N	___ West Indian Locust, *Hymenaea courbaril*		
XN	___ Woman's Tongue, *Albizia lebbek*		
N	___ Divi Divi, *Caesalpinia coriaria*		
N	___ Fish Poison Tree, *Piscidia carthagenesis*		
N	___ Wattapama, *Sabinea florida*		

(Fabaceae cont.)

X ___ Tonka bean, *Diperyx odorata*
X ___ Sea Bean, *Mucuna sloanei*
___ _____
___ _____

Oxalis Family (Oxalidaceae) ___
X ___ Carambola, *Averrhoa carambola*
___ _____

Caltrop Family (Zygophyllaceae) ___
N ___ Lignum Vitae, *Guaiacum officinale*
___ _____

Citrus Family (Rutaceae) ___
X ___ Orange, *Citrus sinensis*
X ___ Grapefruit, *Citrus X Paradisi*
X ___ Lemon, *Citrus limon*
X ___ Lime, *Citrus aurantifolia*
X ___ Tangerine, *Citrus reticulata*
X ___ Other Citrus Trees, *Citrus* spp.
N ___ Prickly Ashes, *Zanthoxylum* spp.
___ Ortanique, hybrid of orange and tangerine
___ _____

Bursera Family (Burseraceae) ___
N ___ Turpentine Tree or Gumbo Limbo, *Bursera simaruba*
___ _____

Trees

Mahogany Family (Meliaceae) ___
XN ___ Mahogany, *Swietenia mahagoni*
X ___ China Berry Tree, *Melia azedarach*
X ___ Neem Tree, *Azadirachta indica*
N ___ Cigar Box Cedar, *Cedrela odorata*
___ _____

Malpighia Family (Malpighiaceae) ___
N ___ West Indian Cherry Tree, *Malpighia emarginata*
N ___ Barbados Cherry, *Malpighia glabra*
N ___ Hogberry, *Byrsonima coriacea*
___ _____

Spurges (Euphorbiaceae) ___
N ___ Manchineel, *Hippomane mancinella*
N ___ Sandbox Tree, *Hura crepitans*
X ___ Gooseberry Tree, *Phyllanthus acidus*
X ___ Monkey-No-Climb, *Euphorbia lactea*
___ _____

Cashew Family (Anacardiaceae) ___
X ___ Mango, *Mangifera indica*
N ___ Cashew Nut Tree, *Anacardium occidentale*
N ___ Hog Plum, *Spondias mombin*

(Anacardiaceae cont.)
N ___ Red Plum, *Spondias purpurea*
N ___ Poison Ash, *Comocladia dodonaea*
___ _____

Soapberry Family (Sapindaceae) ___
X ___ Genip, *Melicoccus bijugatus*
X ___ Akee, *Blighia sapida*
___ _____

Staff Tree Family (Celastraceae) ___
N ___ Nothing Nut Tree, *Cassine xylocarpa*
___ _____

Buckthorns (Rhamnaceae) ___
N ___ Ironwood, *Krugiodendron ferreum*
___ _____

Mallows (Malvaceae) ___
N ___ Tree Hibiscus or Mahoe, *Hibiscus elatus*
X ___ Seaside Hibiscus or Seaside Maho, *Thespesia populnea*
___ _____

Bombax Family (Bombacaceae) ___
XN ___ Silk Cotton or Kapok Tree, *Ceiba pentandra*
X ___ Floss Silk Tree, *Chorisia speciosa*
N ___ Balsa, *Ochroma pyramidale*

(Bombacaceae cont.)
X ___ Guiana Chestnut, *Pachira aquatica*
X ___ Shaving Brush Tree, *Pseudobombax elliptica*
X ___ Baobab Tree, *Adansonia digitata*
X ___ Bombax Trees, *Bombax* spp.

___ _____

Chocolate Family (Sterculiaceae) ___
X ___ Cocoa Bean Tree, *Theobroma cacao*
X ___ Cola Nut Tree, *Cola acuminata*
X ___ Dombeya or Hydrangea Tree, *Dombeya wallichii*
N ___ West Indian Elm, *Guazuma ulmifolia*

___ _____

Annatos (Bixaceae) ___
N ___ Anatto, *Bixa orellano*

___ _____

Cochlospermum Family (Cochlospermaceae) ___
X ___ Brazilian Rose, *Cochlospermum vitifolium*

___ _____

Papayas (Caricaceae) ___
XN ___ Papaya, *Carica papaya*

___ _____

Loosestrife Family (Lythraceae) ___
X ___ Queen Crepe Myrtle, *Lagerstroemia speciosa*

Trees 249

(Lythraceae cont.)
X ___ Henna, *Lawsonia inermis*

___ _____

Pomegranates (Punicaceae) ___
X ___ Pomegranate Tree, *Punica granatum*

___ _____

Brazil Nut Family (Lecythidaceae) ___
X ___ Cannonball Tree, *Couroupita guianensis*
X ___ Barringtonia, *Barringtonia asiatica*

___ _____

Mangroves (Rhizophoraceae) ___
N ___ Red Mangrove, *Rhizophora mangle*

___ _____

Combretum Family (Combretaceae) ___
N ___ White Mangrove, *Laguncularia racemosa*
N ___ Buttonwood, *Conocarpus erectus*
X ___ Tropical or Seaside Almond, *Terminalia catappa*
N ___ Black Olive or Gre-Gre Tree, *Bucida buceras*

___ _____

Myrtle Family (Myrtaceae) ___
X ___ Malay Apple, *Syzygium malaccensis*
XN ___ Rose Apple, *Syzygium jambos*

Trees

(Myrtaceae cont.)
- X ___ Bottle Brush Tree, *Callistemon citrinus*
- N ___ Bay Rum Tree, *Pimenta racemosa*
- N ___ Allspice, *Pimenta diocea*
- N ___ Guava Tree, *Psidium guajava*
- X ___ Eucalyptus, *Eucalyptus robusta*
- XN ___ Surinam Cherry, *Eugenia uniflora*
- N ___ Birch Berry, *Eugenia ligustrina*
- N ___ Other Eugenias, *Eugenia* spp.
- N ___ Guava Berry, *Myrciaria floribunda*

___ _____

Melastome Family (Melastomataceae) ___
- N ___ Kre-Kre, *Tetrazygia eleagnoides*

___ _____

Ginseng Family (Araliaceae) ___
- X ___ Octopus Tree, *Brassaia actinophylla*
- X ___ Aralias, *Polyscias* spp.

___ _____

Theophrasta Family (Theophrastaceae) ___
- N ___ Barbasco, *Jacquinea arborea*
- N ___ Jacquinias, *Jacquinea* spp.

___ _____

Sapodilla Family (Sapotaceae) ___
- N ___ Star Apple, *Chrysophyllum cainito*
- X ___ Naseberry or Sapodilla, *Manilkara zapota*
- N ___ Choky Apple, *Pouteria multiflora*
- N ___ Pan Mango, *Micropholus chryophylloides*
- N ___ Willow Bustic, *Bumelia salicifolia*

___ _____

Dogbane Family (Apocynaceae) ___
- X ___ Red Frangipani, *Plumeria rubra*
- N ___ White Frangipani, *Plumeria alba*
- N ___ Milk Tree, *Tabernaemontana citrifolia*
- XN ___ Lucky Nut, *Thevetia peruviana*

___ _____

Borage Family (Boraginaceae) ___
- N ___ Geiger Tree, *Cordia sebestena*
- N ___ Manjack, *Cordia collococca*
- X ___ Other Cordias, *Cordia* spp.
- N ___ Pigeon Berry, *Bourreria succulenta*

___ _____

Verbenas (Verbenaceae)
- N ___ Fiddlewood, *Citharexylum fruticosum*

(*Verbenaceae cont.*)
X ___ Teak, *Tectona grandis*
___ _____

Black Mangroves
   (Avicenniaceae) ___
N ___ Black Mangrove,
       Avicennia germinans

Bignonia Family (Bignoniaceae) ___
X ___ African Tulip Tree,
       Spathodea campanulata
N ___ Ginger Thomas Tree,
       Tecoma stans
X ___ Jacaranda, *Jacaranda mimosifolia*
X ___ Pink Poui, *Tabebuia rosea*
N ___ Yellow Poui, *Tabebuia serratifolia*
N ___ White Cedar, *Tabebuia heterophylla*
X&N ___ Other Poui Trees, *Tabebuia* spp.

(*Bignoniaceae cont.*)
X ___ Sausage Tree, *Kigelia pinnata*
N ___ Calabash, *Crescentia cujete*
N ___ Black Calabash, *Enallagma latifolia*
X ___ Candle Tree, *Parmentiera cereifera*
___ _____
___ _____

Madder Family (Rubiaceae) ___
X ___ Coffee, *Coffea arabica*
X ___ Ink Berry Tree, *Randia aculeata*
XN ___ Painkiller, *Morinda citrifolia*
N ___ Wild Coffee, *Psychotria nervosa*
___ _____
___ _____

Pages for Added Observations